Springer Proceedings in Physics 308

Indexed by Scopus

The series Springer Proceedings in Physics, founded in 1984, is devoted to timely reports of state-of-the-art developments in physics and related sciences. Typically based on material presented at conferences, workshops and similar scientific meetings, volumes published in this series will constitute a comprehensive up to date source of reference on a field or subfield of relevance in contemporary physics. Proposals must include the following:

- Name, place and date of the scientific meeting
- A link to the committees (local organization, international advisors etc.)
- Scientific description of the meeting
- List of invited/plenary speakers
- An estimate of the planned proceedings book parameters (number of pages/articles, requested number of bulk copies, submission deadline).

Please contact:

For Americas and Europe: Dr. Zachary Evenson; zachary.evenson@springer.com
For Asia, Australia and New Zealand: Dr. Loyola DSilva; loyola.dsilva@springer.com

Abhishek Tiwari · Pratik Kumar Ray ·
Neha Sardana · Rajiv Kumar
Editors

Proceedings of the International Conference on Fundamental and Industrial Research on Materials

iConFIRM 2023, 11–14 December, Ropar, India

Springer

Editors
Abhishek Tiwari
Metallurgical and Materials Engineering
Indian Institute of Technology Ropar
Rupnagar, Ropar, Punjab, India

Pratik Kumar Ray
Metallurgical and Materials Engineering
Indian Institute of Technology Ropar
Rupnagar, Ropar, Punjab, India

Neha Sardana
Metallurgical and Materials Engineering
Indian Institute of Technology Ropar
Rupnagar, Ropar, Punjab, India

Rajiv Kumar
Metallurgical and Materials Engineering
Indian Institute of Technology Ropar
Rupnagar, Ropar, Punjab, India

ISSN 0930-8989 ISSN 1867-4941 (electronic)
Springer Proceedings in Physics
ISBN 978-981-97-4556-2 ISBN 978-981-97-4557-9 (eBook)
https://doi.org/10.1007/978-981-97-4557-9

This Springer imprint is published by the registered company Springer Nature Singapore Pte Ltd.
The registered company address is: 152 Beach Road, #21-01/04 Gateway East, Singapore 189721, Singapore

Preface

The International Conference on Fundamental and Industrial Research on Materials (iConFIRM 2023) was held at IIT Ropar from 11–14 December 2023. The four-day extravaganza started with a workshop on damage modeling using cohesive zone and user subroutines (sponsored by Dassault System) and X-ray diffraction characterization technique for materials science (sponsored by Malvern Panalytical) on 11 December 2023, followed by three exciting days of research talks from 12–14 December 2023. The conference included exciting plenary talks from all over the world and promoted greater intellectual interactions between faculty, scientists, students, and industry. The conference also provided a common platform for a faculty/scientist in search of good graduate students/post-docs, a student in search of academic/industrial research jobs, and an industrialist on the lookout for talent to solve complex problems. Revolutionary approaches opened avenues for accelerated materials research, development, and manufacturing. This conference aims to bring together leading researchers, students, and industry under one umbrella to foster meaningful discussion and collaboration.

The present volume of proceedings contains full research papers covering a wide spectrum of topics within the field of metallurgy. From novel extraction and recycling techniques to the development of advanced structural materials, from understanding corrosion/oxidation mechanisms to exploring the potential of functional materials, the contributions presented in the conference reflect the depth and breadth of worldwide research being conducted. These areas represent not only the core pillars of metallurgical research but also the driving forces behind innovation and sustainability in various industries. After receiving the full papers from various speakers, the papers were sent for a rigorous review process. About 19 research papers were accepted after a rigorous peer-review process. The accepted papers represent a set of high-quality research contributions in the field of materials.

As editors of the conference proceedings, we extend our gratitude to all the authors who have contributed their work to this volume. We are immensely grateful to all the authors who contributed their research papers, posters, and presentations to the conference "iConFIRM 2023". We are also grateful to the reviewers whose dedication, expertise, and passion have enriched this volume. Additionally, we would like to thank the organizing committee, sponsors, and participants whose dedication and enthusiasm made the "iConFIRM 2023" conference a success (Fig. 1).

Fig. 1. Top Let panel: Inauguration of the conference; Poster session; Poster session; Plenary talk

We hope that the proceedings of the "iConFIRM 2023" conference will serve as a valuable resource for researchers, practitioners, and students alike, inspiring further exploration and innovation in the fascinating field of metallurgy.

April 2024

Pratik Kumar Ray
Neha Sardana
Abhishek Tiwari
Rajiv Kumar

Organization

Core Organizing Committee

Patron

Rajeev Ahuja	IIT Ropar

Chair

Pratik K. Ray	IIT Ropar

Convener

Neha Sardana	IIT Ropar

Members

Ravi Mohan Prasad	IIT Ropar
Khushboo Rakha	IIT Ropar
Abhishek Tiwari	IIT Ropar
Atharva Poundarik	IIT Ropar
Prince Kumar Singh	IIT Ropar
Rajiv Kumar	IIT Ropar

Advisory Board Members

B. P. Kashyap	IIT Jodhpur
Manoj Gupta	NUS Singapore
Prita Pant	IIT Bombay
Rajesh Prasad	IIT Delhi
R. Jayaganthan	IIT Madras
S. Seetharaman	KTH Sweden
Mufit Akinc	Iowa State University

Sponsors

Material Recycling Association of India
Dassault Systems
JEOL India Pvt. Ltd.
Aarti Steel International Limited
Spectris Technology Pvt. Ltd.

Contents

Extractive Metallurgy and Recycling of Materials

Fluidized Bed Roasting of Zinc Sulfide Concentrate: Role of Particle Size

Suvam Mukherjee(✉)

Central Research and Development Laboratory, Hindustan Zinc Ltd., Udaipur 313024, Rajasthan, India
Suvam.Mukherjee@vedanta.co.in

Abstract. Industrial roasting of zinc sulfides into its oxide/calcines is customarily practiced in fluidized bed reactor. High extent of conversion is desired to facilitate acid-leaching of the zinc calcine. Industrially, degree of such roasting is quantified in terms of the sulfide-sulfur index describing extent of sulfides being unreacted. This is governed by multiple roasting process parameters (*viz.*, concentrate feed rate, air flow, moisture etc.) and essentially particle residence time of the feed concentrate inside roaster.

In the present study, an attempt has been made to estimate the roasting process from analysis of calcines obtained from different sites of roaster, especially with regards to the role of particle size. Towards this, industrial sample characterization, minerology and plant data monitoring has been done. Results of such, indicated the calcine size ranges where extent of roasting is minimal. Furthermore, possibilities of precious metal recoveries from calcines after being withdrawn from specific roaster locations has also been outlined that effectively lacked in past investigations. Finally, owing to the complexity of such high temperature process, need for improving process monitoring techniques for superior process stabilization has also been recommended.

Keywords: Zinc sulfide · fluidised bed reactor · calcine

1 Introduction

Fluidized bed roasting of sulfide ore concentrates for producing zinc has been the popular industrial route since the invention of Lurgi based FBR [1] way back in 90s. It marks an essential pyrometallurgical step wherein the conversion process is aided by fluidization in presence of an oxidizing environment [2]. The product thus formed act as feed/input for the rest downstream process i.e., leaching followed by electrowinning (EW) as per R-L-E based zinc circuit. A brief schematic of the roasting circuit is given in Fig. 1 for reader's understanding. Detailed description of the RLE process is well known in literature [3], thus not re-iterated here. Looking into the overall process of Zinc-making, the roasting step entitles the vital yet complex operation that decides the quality of downstream processes. Consequently, researchers [2, 4] over the past decade worked on roasting process optimization by virtue of experiments along with soft computing that controls process parameters with an overall aim of optimizing roasting efficiency.

A. Tiwari et al. (Eds.): iConFIRM 2023, SPPHY 308, pp. 3–13, 2024.
https://doi.org/10.1007/978-981-97-4557-9_1

Although such, the particle size, its residence time inside roaster and impact on roasting efficiency has been overlooked inspite of its equivalent importance. The present study focuses on such aspect to estimate the roasting efficiency along with possible proposals for essential minor-metal recoveries that no-doubt has been unexplored in past literatures.

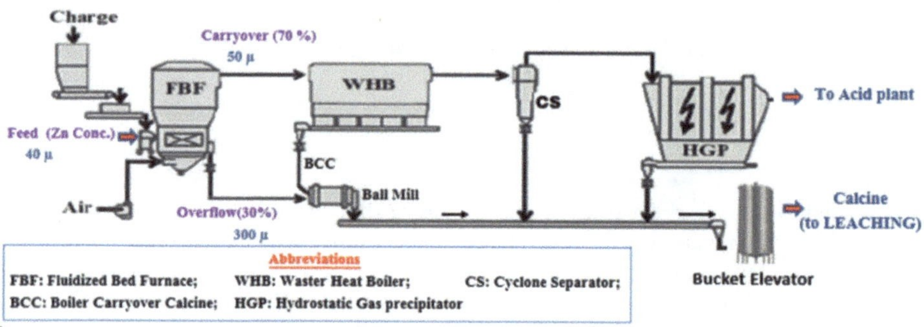

Fig. 1. Schematic representation of an industrial zinc sulfide roasting circuit prior to leaching

2 Present Study

As the first attempt to study the impact of particle size in the extent of ZnS roasting, calcine samples from different locations of industrial roaster were collected, namely.

Roaster top and the calcine collected from the roaster top is the *carryover calcine.* Since BCC calcine holds the maximum share of overall carryover calcine, thus in the present study, BCC calcine analysis has only been considered as representative of overall carryover calcine.
Roaster Bottom (bed) and the calcine collected from the roaster bottom or bed is the *overflow calcine*
Bucket elevator (BE): The *BE calcine* in this location is the final roaster calcine, mainly a mixture of calcines collected from both top and bottom of roaster. This final calcine goes for successive leaching process.

The aforementioned locations have been schematically demonstrated in Fig. 1, for reader's ease of understandability. To anticipate roasting process efficiency, customarily time varying calcines from these locations are collected and sent for respective chemical analysis so as to estimate the degree of unroasted in calcine, better termed as Sulfide-sulfur (SS) index. Accordingly, transient SS variation of these calcines is demonstrated in Fig. 2(a) wherein it is observed that among the two different calcine collections points, carryover calcine have higher SS compared to the overflow. Moreover, the trend for BE calcine SS explicitly suggest the superior contribution of carryover calcine in deciding overall roaster calcine SS. Reasonable calculation has been made in Sub-sect. 3.1. An explanation to this differential calcine SS behavior is due to varied particle residence time inside the roaster. Fine calcines collected from the roaster top owing to its lower residence time, have less chances of roasting resulting in its high SS in calcine. Past works of literatures [2] confirms detailed discussion on the same, thus not re-iterated here. Prior to discussion of the results in the present study, a tabulated representation of the roaster process parameters has been made in Table 1 for reader's ease of understanding.

Table 1. Relevant roaster operating conditions in a 1.2 MMTPA plant

Sl. No.	Process Parameters	Values
1	Zinc concentrate feed rate, TpH	42.1
2	Total air flow rate, Nm^3/hr	75447
3	Secondary air flow rate, Nm^3/hr	36216
4	*% of primary air as Sec air*	*48%*
5	Roaster top temperature, °C	945
6	Roaster water injection, m^3/hr	5137
7	Moisture in feed concentrate	10.93%
8	Average roaster bed pressure, mmWC	1627
9	Average roaster bed temperature, °C	858

3 Results and Discussions

3.1 Predicting Contribution of Different Location Calcines Towards Overall Calcine Sulfide-Sulfur Outcome

As per the general mass balance calculation,

$$\left(f_{Carryover} \times SS_{Carryover}\right) + \left(f_{Overflow} \times SS_{Overflow}\right) = SS_{Bucket\ Elevator} \tag{1}$$

Referring to Fig. 2(a), average value can be accordingly taken as

$$\begin{aligned} &SS_{Overflow} = 0.2;\ SS_{Carryover} = 0.6;\ SS_{Bucket\ Elevator} = 0.5 \\ &\text{and furthermore, from mass balance } f_{Carryover} + f_{Overflow} = 1 \end{aligned} \tag{2}$$

Therefore, Eq. (1) can be arranged and re-written as,

$$\left(f_{Carryover} \times 0.4\right) = 0.3 \tag{3}$$

thus, $f_{Carryover} = 0.75$

where, f and SS are mass fraction and sulfide-sulfur of respective calcines. Referring to above calculation, as per the individual calcine sulfide-sulfur index, it is well established that roughly 70% of total bucket elevator calcine comes from the roaster top and the rest 30% from the roaster bed. This can also be explained from the calcine particle size variation. Referring to the results of experimental analysis in Fig. 2(b), carryover calcines has fine size range with p80 of ~60 μ whereas overflow/bed calcine have the p80 of ~350 μ. Owing to fine size range of carryover calcine, they experience less residence time inside the roaster leading to lesser chance of complete roasting of Zn sulfide thus, resulting in its high SS. On the other hand, overflow calcine by virtue of its higher residence time gets ample time to roast and converts to Zn oxides, subsequently having low calcine SS.

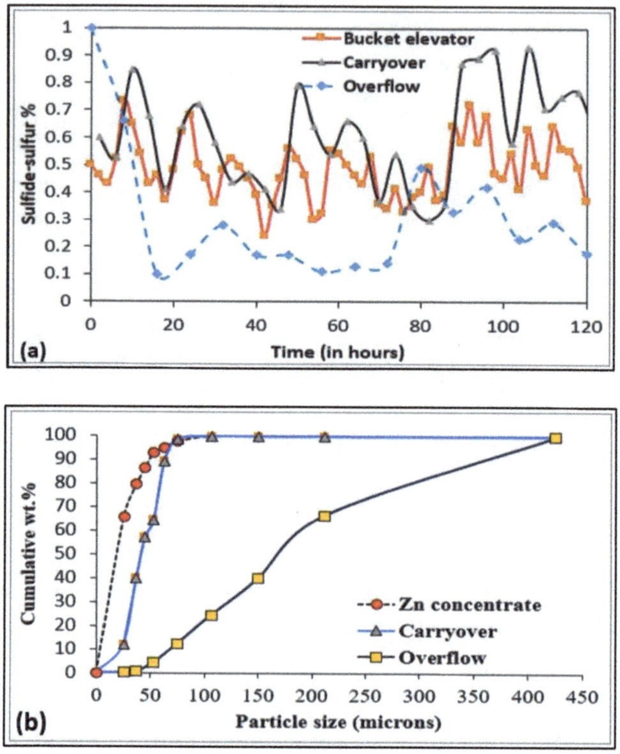

Fig. 2. Variation of calcine (a) sulfide-sulfur (SS); (b) particle size distribution from different roaster locations collected during industrial trials

3.2 Minerology Results of Calcines Collected from Different Roaster Locations

Collected samples from roaster i.e., both zinc concentrate and calcines have been characterized in FESEM [specified as Carl Zeiss (Sigma 300) with Bruker EDS (XFlash 6160)], as facilitated by the Central R&D Lab. of Hindustan Zinc Ltd. Micrographs of the respective samples have been demonstrated in Fig. 3. There, a lucid difference in average particle size between the carryover and the overflow/bed calcine (in Fig. 3(b) and (c)) can be distinguished. Interestingly, particle size of concentrate that is fed in roaster and carryover seems visually identical as one can refer from Fig. 3(a) and (b) respectively. This, as also confirmed from experiment, p80 [5] of both the calcines (see Fig. 2(b)) is <50 μ. This observation is indicative to the fact that at the practiced bottom airflow in roaster, the finer range of the feed directly vanishes as carryover after being fed inside. On the other hand, polished micrograph of bed calcine (Fig. 3(c) exclusively shows the agglomeration [6] of particles happening over the course of feed roasting. Finally, the overall calcine (as shown in Fig. 3(d)) i.e., the mixture of calcines collected from roaster top and bottom, going for leaching has the p80 < 60 μ after reasonable ball milling of bed calcines.

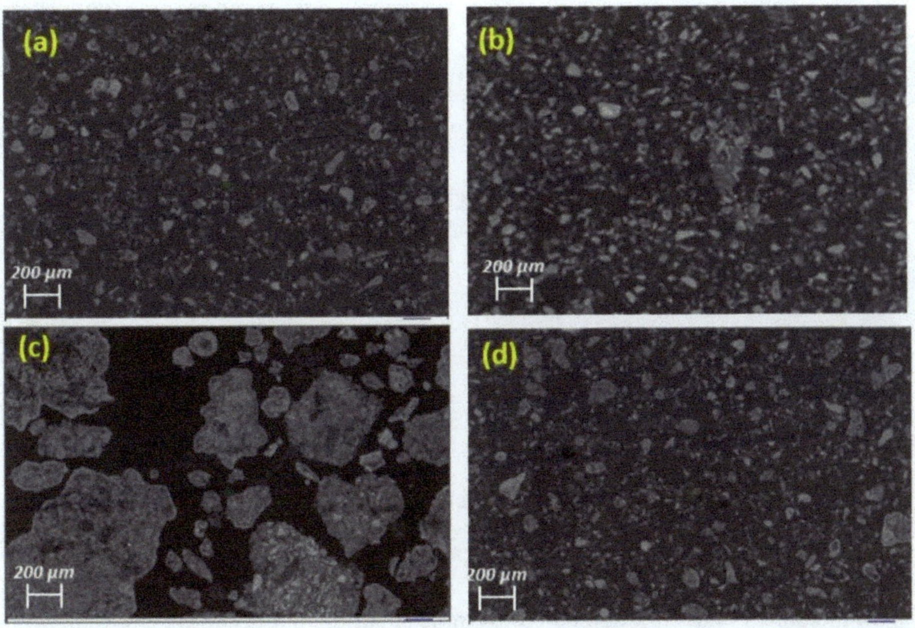

Fig. 3. Back-scattered electron micrographs of polished samples of (a) zinc concentrate [roaster feed]; (b) BCC calcine; (c) overflow [bed] calcine and (d) bucket elevator calcine, respectively.

3.3 Comparison of Calcine Particle Sizes with Degrees of Roasting

The collected calcines from two locations of the roaster successively gone for SS test and particle size analysis. The results of the same has been demonstrated in Fig. 4 and alongside, following observations can be outlined:

1. High calcine SS is observed in both the finest and coarsest range of the parent calcine.
2. The calcine whose overall SS is high, its SS at any particular size range also reflects superiority when compared with calcine of lower overall SS.
3. Contradictory to observation (2), no such variation in the calcine particle size observed when talking of overall high/low calcine SS.
4. Same SS in ~ 150 μ particles in carryover and overflow hints at good transfer of such particles between bed and freeboard.

As per observation (1), high SS can be found in both carryover and overflow calcines. In case of carryover calcines, finer fractions have a lower particle residence time inside roaster and thus remain unroasted (as also mentioned in sub-Sect. 3.1) ultimately contribute to high SS. This explanation corroborates with the micrograph in Fig. 5(a). There as seen, unreacted ZnS particle is prominent surrounded by minimal oxide layer and the thickness of this oxide layer rely on residence time and oxygen concentration inside roaster.

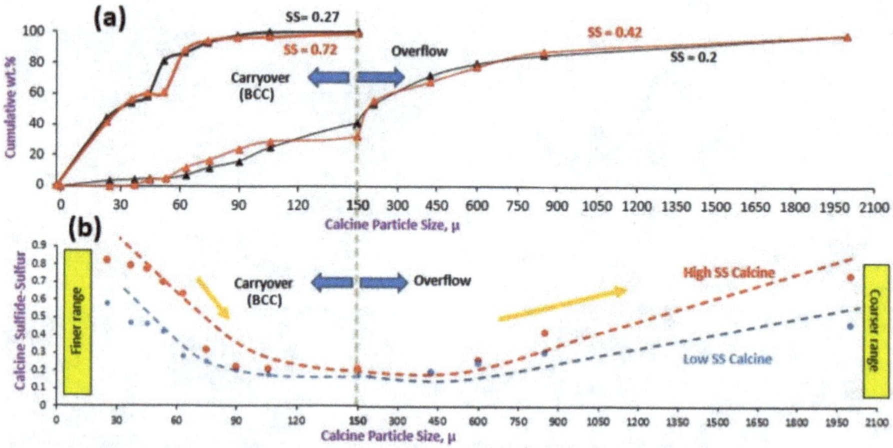

Fig. 4. Variation of calcine particle size with its (a) respective weights and (b) sulfide-sulfur in BCC and overflow/ bed calcines respectively

On the other hand, although an unfavorable situation, but occurrence of high SS in overflow/bed calcine in its coarsest section (as seen in Fig. 4) can be due to unavoidable particle agglomeration of the bed calcine during the roasting regime. Individual ZnS particles get adhered to each other in presence of fused impurities like silicates and forms a partially roasted agglomerate (see Fig. 5(b)). Detailed understanding and occurrence of calcine agglomeration during roasting is well document in past literature [6], thus not re-iterated here. The formation of agglomerated calcines with silicious boundaries provide a hindrance to oxygen diffusion inside the particle, thereby leading to formation of unroasted core with oxidized surfaces. Detailed diffusion kinetics is already mentioned in past [7, 8]. Further, presence of an unroasted core with surface of low melting point impurities viz., silicates is seen in Fig. 5(b[i]).

Thus, micrographs with chemical analysis results explain that in both size extremities, partial or incomplete roasting of ZnS is leading to higher SS in those ranges, however, causes of such are different. While fine unroasted calcines having terminal velocity lower than air velocity goes from the top and the same with velocity greater than air velocity accumulates in bed.

Although such, a good fraction of bed calcine consists of fully roasted agglomerate (as shown in Fig. 5(c)) essentially embodying zincite (i.e., zinc oxide) with traces of ferrites (Zn-Fe oxides) as seen in phase distribution micrograph in Fig. 5(c[i]).

Now, the unroasted calcines thus formed due to differential particle interaction with oxidizing environment, such are very detrimental to the downstream leaching process as explained by the reaction in Eq. (4). The presence of sulfides in calcine actually reverts back the ferric to ferrous ions in leaching stages leading to decrease in solid-liquid separation. Detailed explanation to this is beyond present study's context.

$$ZnS + 2\,Fe^{3+} \rightarrow 2\,Fe^{2+} + \text{Sulfur} \tag{4}$$

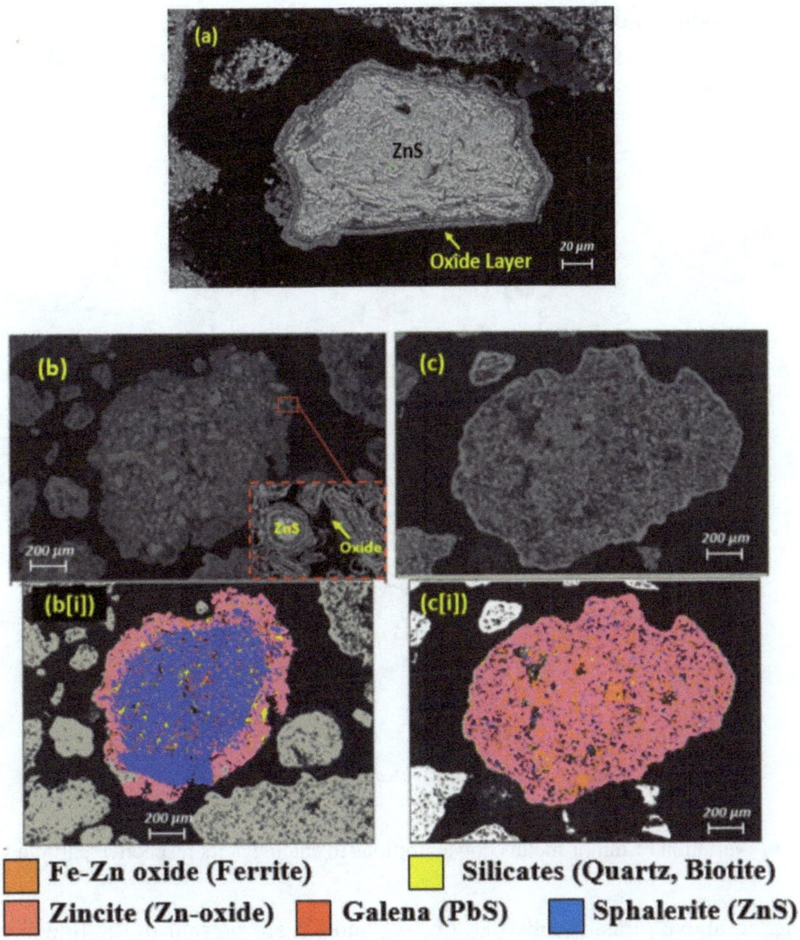

Fig. 5. Back-scattered electron micrograph of (a) partially roasted single ZnS particle in BCC calcine; (b) partially and (c) completely roasted agglomerated particle in overflow/bed calcine. {b[i] and c[i] shows various phase mapping within a grain as given in (b) and (c)}

3.4 Possibilities of Valuable Metal Recovery from Calcines Post Roasting

As a general industrial practice, calcine produced from fluidized bed roasting is stored in silos (i.e., storage vessel) and later on goes for the leaching process wherein the Zn-oxides (i.e., calcine) are treated with H_2SO_4 and respective sulphates are produced. Researchers are working on recovery of minor metals *viz.*, Ag, Cu, Pb from the residues formed during the leaching stages whereas, no such treatment process between roasting and leaching is presently practiced in industries. Therefore, keeping in mind of the economic aspects of their recoveries, attempt have been made in the present work so as to explore whether possibilities of valuable metal recovery, if possible, post roasting. To this end, collected calcines have been segregated in different size ranges and elemental analysis via AAS (Atomic absorption spectroscopy) of calcines of these individual size ranges have been done. Results of such is given in Fig. 6.

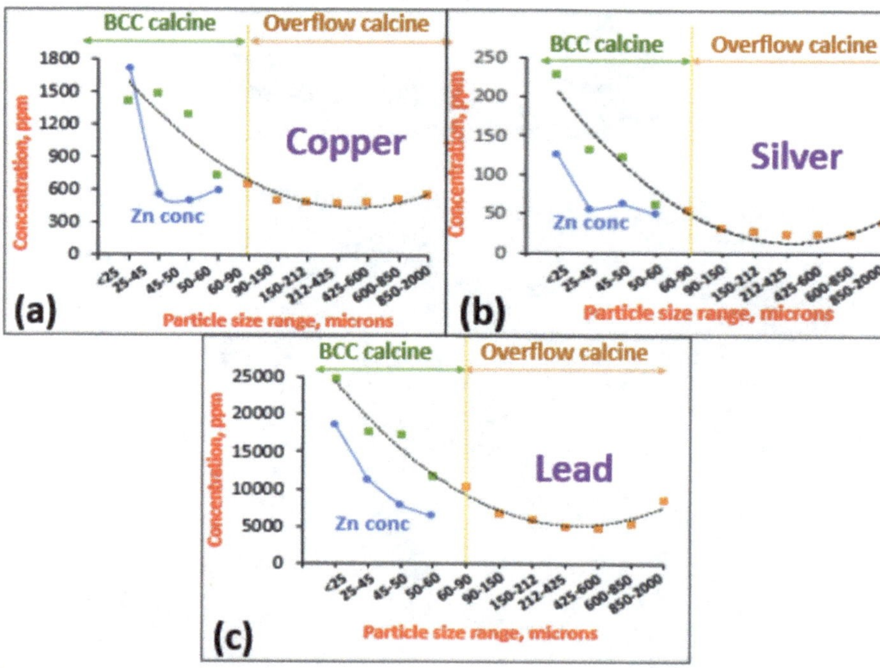

Fig. 6. Variation of minor metals concentrations in calcine at its respective particle sizes

Refer to above plots, metals like Pb, Ag and Cu are present in the fine size range in the Zn concentrate (roaster feed) and thus seen most in the BCC calcines i.e., part of the carryover calcine. However, these metal concentrations subsequently decrease as particle size increases. Among these, silver or 'Ag' being a precious metal, its abundant presence in finer range (see in Fig. 6(b)) is a positive indication in terms of scopes of metal recovery. Quantitatively, around half of Ag in Zn concentrate goes to 15% of the fines (<25 μ, Ag ~ 250 μ ppm) of carryover. Herewith, a process can be advanced consisting of separating calcines ≤30 μ particle size by means of industrial sieving or alternatively, stream diverting the HGP calcines (see HGP location in Fig. 1) where, Ag ~ 300 ppm, from the roaster circuitry, for recovery process. By doing so, calcine availability for leaching step also will not be affected as much since HGP calcine contributes to ~7–10% of total calcine produced during roasting. Having investigated these, currently possible economic route of Ag recovery in fine calcines prior to conventional leaching is under process of exploration.

Alongside, presence of high Pb in fine calcine range as seen in Fig. 6(b) suggests the presence of unburnt Pb-sulfides coming from the feed concentrate. Fig. To confirm the above, SEM characterization of BCC [part of carryover calcine] is done as shown in Fig. 7(a) below. There, along with common ZnS phase, white regions denoting the Pb rich sulfide phase in the grain are present in abundance and this contributes to an increase in calcine SS as also seen in the plot in Fig. 7(b).

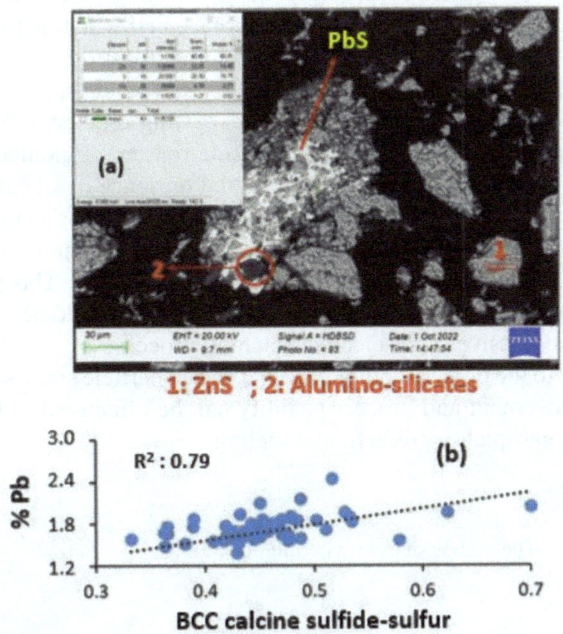

Fig. 7. (a) FESEM micrograph of BCC calcine showing unreacted PbS, ZnS and Alumino-silicate phases; (b) showing variation of Pb concentration with BCC calcine sulfide-sulfur.

3.5 Potential Research Opportunities

Generally high temperature metal extraction process has its complexities and thus, maintaining process with lowered fluctuations is distantly achievable since the process is multi-parameters dependent. In our present study, it is seen that feed particle size plays a vital role in deciding roasting efficiency viz., calcine SS. One solution to such is feed pelletization [8] prior to roasting but not feasible for every industry due to cost limitations and high production rate. In this context, process efficiency prediction via soft sensing and data predictive modeling are viable alternatives. Towards this, author is exploring opportunities of roasting efficiency (calcine SS) prediction by means of off gas analysis and data modeling of existing process parameters.

4 Conclusions

From the present study on the impact of particle size on the roasting efficiency of zinc sulfide in a fluidized based Lurgi roaster, following conclusions can be drawn:

- During roasting, the calcines collected from top and bottom locations have distinctly different average size ranges. Chiefly, having p80 of 50 and 300 μ, respectively. Owing to such differential average calcine size range, residence time of the particles inside the roaster also varies accordingly that affects the roasting efficiency.
- Coarser sized calcine is collected from the furnace bed which is the outcome of the agglomeration process occurring during concentrate roasting.
- Calcine in the finest size range (in case of carryover) and the coarsest range (from bed) experience high sulfide-sulfur i.e., poor roasting efficiency. While the former is due to minimal particle residence time inside roaster, agglomeration of unburnt ZnS and thus, lowered O_2 diffusion owing to the presence of silicates and other low melting point impurities in bed material explains poor roasting in the latter situation.
- Interestingly, in carryover section, calcines with size <30 μ have possibilities of precious metal recoveries essentially silver, copper and lead. This provides a scope of especially Ag recovery by by-passing 15% of overall calcine and en-routing to screening and successive pyro/selective leaching processes.
- Finally, owning to the process complexity and varied particle sizes, scopes of roasting efficiency improvement and process stability can be effectively achievable via soft sensing and off gas quality prediction modeling.

Conflict of Interest. The author shares no conflict of interest.

References

1. Choi, C.-Y., Chun, Y.-G.: Roaster with stabilized fluidized bed for roasting zinc concentrate. Korea Zinc Co. Ltd., US patent: 6669894B2, pp. 1–8 (2001)
2. Saxen, B., Nyberg, J, Harkonen, J., Toivonen, H.: Dynamic behaviour of a zinc roaster furnace. In: Automation in Mining, Minerals and Metal Processing, pp. 209–214 (2004)
3. Sinclair, R.J.: The extractive metallurgy of Zinc. 1st edn. AusIIM, Australia (2005)
4. Feng, Z., Li, Y., Sun, B., Yang, C., Zhu, H., Chen, Z.: A trend-based event-triggering fuzzy controller for the stabilizing control of a large-scale zinc roaster. J. Process. Control. **97**, 59–71 (2021)
5. Bond, F.C.: Crushing and grinding calculations Part 1. Br. Chem. Eng. **6**(6), 378–385 (1961)
6. Benlyamani, M., Ajersch, F.: Agglomeration of particles during roasting of Zinc sulfide concentrates. Metall. Trans. B **17B**, 647–686 (1986)

7. Chen, T.T., Dutrizac, J.E.: Minerological changes occurring during the Fluid-bed roasting of zinc sulfide concentrates. JOM **56**, 46–51 (2004)
8. Heukelman, S., Groot, D.: Fluidized bed roasting of micro pelletized zinc concentrate: Part I-Pellet strength and roasting kinetics. J. S. Afr. Inst. Min. Metall. **111**, 759–765 (2011)

Investigation on Solid-State Carbothermic Reduction Behaviors of Pyrolusite Ore

Sambit Tripathy[1] and Navneet Singh Randhawa[2(✉)]

[1] Department Metallurgical and Materials Engineering, Indian Institute of Technology Ropar, Rupnagar 140001, Punjab, India
[2] Metal Extraction and Recycling Division, CSIR-National Metallurgical Laboratory, Jamshedpur 831007, Jharkhand, India
nsr@nmlindia.org

Abstract. Manganese is essential to iron and steel production because of its sulfur-fixing, deoxidising, and alloying properties. Also, manganese has no satisfactory substitute in producing aluminium alloys and battery cells. It was estimated that manganese ore consumption reached about 200 million metric tons in the last decade. The most important manganese ore is pyrolusite (MnO_2), which is composed of higher manganese oxide and needs more energy to obtain manganese metal. Previous studies suggest that pre-reduced ore pellets in FeMn production lead to a significant drop from ~3000 to ~1600 kWh t^{-1}in energy consumption. Hence, the prereduction of manganese ores is essential to increasing ferromanganese production efficiency and cost-effectiveness. This study analysed the kinetics and mechanism of roasting of pyrolusite manganese ore for use in ferromanganese production. Mixtures of ore-coke in a ratio of 1:0.15 were prepared for carbothermic reduction. The effect of ore sizes (5–20 mm), reduction temperatures (900–1100 °C) and holding time (60–180 min) on the reduction behaviour were examined. Optimum conditions were obtained for a higher reduction of ore. The decomposition of higher oxide states of manganese was confirmed from XRD analysis of raw and reduced ores. Our study indicated that the reduction of the ore was inversely dependent on its size, whereas a higher temperature and the holding time improved the reduction of the ore.

Keyword: Carbothermic reduction · Roasting: Ferromanganese · Manganese Ore · Kinetics

1 Introduction

Manganese is strategically essential in many fields, including steel production, chemical, paint, and battery industries. Around 95% of the manganese alloys, mainly produced from manganese ore, have been included as deoxidisers, desulfurizers, and alloys in the steelmaking industry [1]. To produce ferromanganese, blast and electric arc furnaces use lump manganese ore, sinter, and metallurgical coke [2]. Rising energy costs and environmental concerns associated with producing sintered ores and metallurgical coke drive the search for alternative technologies. Prereduction of manganese ores

A. Tiwari et al. (Eds.): iConFIRM 2023, SPPHY 308, pp. 14–20, 2024.
https://doi.org/10.1007/978-981-97-4557-9_2

could be an appealing route to increasing ferromanganese production efficiency and cost effectiveness [3].

After being pre-reduced, the material can be melted in an open arc furnace to produce ferromanganese alloy and slag. According to some reports, using pre-reduced pellets in an electric furnace rather than ore caused the drop from ~3000 to ~1600 $kWht^{-1}$ in energy consumption for FeMn production [4, 5]. Manganese is present in manganese ores as higher oxides and silicate, and during the early stages of reduction, these are reduced to MnO by the reduced atmosphere in the furnace [6]. Reducing MnO to carbide and metal is the most critical reaction in producing ferromanganese alloys. It can occur by two mechanisms:(a) Reduction in the solid state by carbon [7, 8], (b) Dissolution of MnO in the slag, followed by reduction from the slag by a solid carbonaceous reducing agent or carbon-saturated alloy.

The former mechanism is the subject of the study reported in this paper, as the latter mechanism has been extensively observed in laboratory-scale experiments [5]. The carbothermic reduction of manganese ores and pure manganese oxides has been the subject of numerous studies [9–11]. Researchers Terayama and Ikeda [11] and Rankin and Deventer [10] explored manganese oxide reduction by graphite at 1070–1425 °C in an inert atmosphere. The reaction of graphite gasification was discovered to be the rate-controlling step in both of these studies. Chemical reactions control the rate of reaction at low temperatures (up to 1000 °C) [13], while diffusion is responsible for the rate at temperatures above 1100 °C [14]. Figure 1 depicts the reactions at various temperatures when a manganese ore and coke mixture is roasted [15–22].

Reactions	ΔG° (kJ/mol)	T (°C)
Reduction reactions of oxygen-rich oxides to MnO		
$3Mn_2O_3 + C = 2Mn_3O_4 + CO$	ΔG° = - 0.25 - 0.17T	25 - 1100
$3Mn_2O_3 + CO = 2Mn_3O_4 + CO_2$	ΔG° = - 170.71 - 0.004T	25 - 1100
$Mn_3O_4 + C = 3MnO + CO$	ΔG° = 110.96 - 0.21T	25 - 1244
	ΔG° = 84.35 - 0.20T	1244 - 1700
$M_3O_4 + CO = 3MnO + CO_2$	ΔG° = 110.96 - 0.21T	25 - 1244
	ΔG° = 84.35 - 0.20T	1244 - 1700
Reduction reaction of MnO with carbon monoxide		
$MnO + CO = Mn + CO_2$	ΔG° = 102.38 + 0.01T	25 - 1227
	ΔG° = 116.73 + 0.01T	1227 - 1727
Boudouard reaction		
$CO_2 + C = 2CO$	ΔG° = 170.82 - 0.18T	25 - 1727
Reduction reactions of MnO with carbon or iron carbide		
$MnO + C = Mn + CO$	ΔG° = 287.6 - 0.16T	25 - 1227
$MnO + 10/7C = 1/7Mn_7C_3 + CO$	ΔG° = 284.22 - 0.18T	717 - 1087
	ΔG° = 282.01 - 0.18T	1087 - 1137
	ΔG° = 280.22 - 0.18T	1137 - 1244
	ΔG° = 280.35 - 0.18T	1244 - 1700
$MnO + 10/7Fe_3C = 1/7Mn_7C_3 + 30/7Fe + CO$	ΔG° = 246.09 - 0.15T	717 - 840
	ΔG° = 269.42 - 0.17T	840 - 1087
	ΔG° = 267.42 - 0.17T	1087 - 1137
	ΔG° = 265.42 - 0.17T	1137 - 1244

Fig. 1. Chemical reactions occurring in the reduction roasting of manganese ore

This study attempted to establish fundamental knowledge of the kinetics and mechanism of manganous oxide reduction by coke against the effect of different ore sizes, reduction temperatures and holding time on the reduction rate. This understanding of the processes and transitions that occur during the reduction of manganese ores while manufacturing ferromanganese alloys should be explored.

2 Experimental Procedure

Raw materials, reduction behaviour and reduced sample characteristics were systematically researched. The experimental procedure can be described as follows: The mixture of coke with manganese ore in a ratio of 0.15 was prepared for carbothermic reduction. Roasting of the mixtures was carried out in a raising hearth furnace. Reduced samples were then cooled to room temperature inside the furnace. The raw materials were dried in an oven chamber at 115 °C for 2 h. The coke used in this investigation was crushed to less than 5 mm in size using a jaw crusher and pulveriser. Additionally, manganese ores with different size ranges (5–10 mm, 10–15 mm, 15–20 mm) were prepared to study particle size's impact on the reduction process. The raw manganese ore of about 1000 g and two parts of coke, each of sizes ranging from less than 5 mm to coke fines, was adequately mixed with raw ore at a ratio of 15% by wt. of the raw material to get a ratio of coke to manganese ore of 0.15.

Next, the mixtures were charged into a graphite crucible and placed into the raising hearth furnace. A layer of coke fines covers the mixed material surface with a proportion of 10% of the initial coke added to the raw material [19]. This reduction experiment looks into the impact of roasting manganese ore at different temperatures and for different amounts of time. After the criteria set for the investigations were met, samples from all the products obtained were prepared separately for each experiment, sieved through a 5 mm sieve, dried, weighed and analysed for Fe. The metallisation ratio of the reduced samples was measured after each roasting experiment. The metallisation ratio (MR) was evaluated with Eq. (1):

$$MR = M_{Fe} / T_{Fe} \times 100\% \tag{1}$$

M_{Fe} was the weight of metallic iron, and T_{Fe} was the weight of total iron in the reduced samples. The commonly used $FeCl_3$-$K_2Cr_2O_7$ volumetric method was applied to M_{Fe} and T_{Fe} measurements.

Apart from the iron analysis of the reduced samples, chemical analysis of initial raw ore and coke was also carried out. Before and after reduction, Ore samples were sent for X-ray diffraction (XRD) studies. XRD studies used a Bruker (Model: D8, Discovery) US machine with Cu-Kα radiation at a 1°/min scan rate. Powder X-ray diffraction (XRD) patterns were recorded in the 2θ range 10–80°. Crystalline phases in the samples were identified using ICDD PDF standards (powder diffraction files).

3 Results and Discussions

The chemical compositions of manganese ore used in this study are in Table 1. The fixed carbon percentage of Coke was found to be 74%. The raw manganese ores were high-grade with manganese content (45.50 wt%) and low Fe content (8.8 wt%) at a high Mn/Fe ratio of 5.17.

Table 1. Composition of elements in manganese ore.

Composition	Mn	Fe	Al_2O_3	Si	CaO	MgO	P
Weight %	45.50	8.8	7.9	1	1	0.37	0.08

XRD analysis of raw and reduced ores shows us the decomposition of higher oxide states of manganese from MnO_2 to Mn_3O_4 and MnO as shown in Figs. 2 and 3. It can also be inferred from the prominent peaks of FeO that iron in the form of Fe_2O_3 has also decomposed after the carbothermic reduction process.

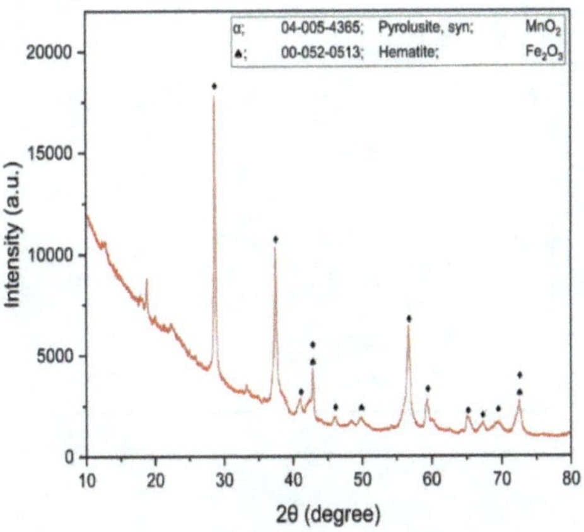

Fig. 2. XRD analysis of Raw Pyrolusite Ore

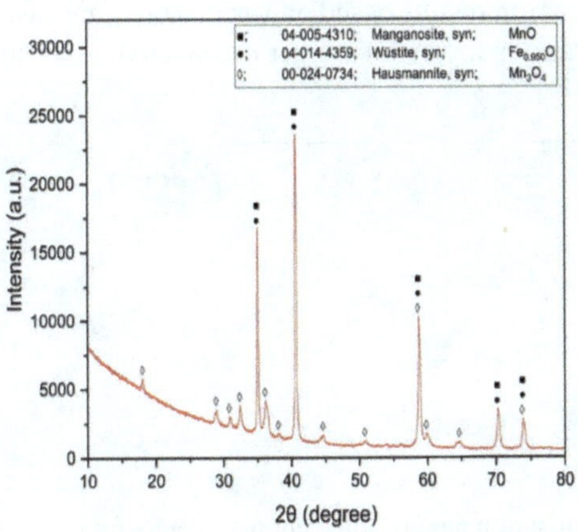

Fig. 3. XRD analysis of Roasted Ore showing decomposition of pyrolusite after reduction

The size of the ore plays a crucial role in the reduction process. By plotting the data for weight loss of materials against controlled conditions set in this study, done in Fig. 4, we can observe that smaller ore sizes have a higher weight loss than their counterparts.

Generally, smaller-sized ores are preferred for reduction because they have a larger surface area per unit mass, facilitating faster and more efficient reduction. This is because the reduction process involves the transfer of heat from the coke to the ore particles. Smaller-sized ores have a larger surface area, which means more sites are available for the coke to react with, leading to a faster reduction rate.

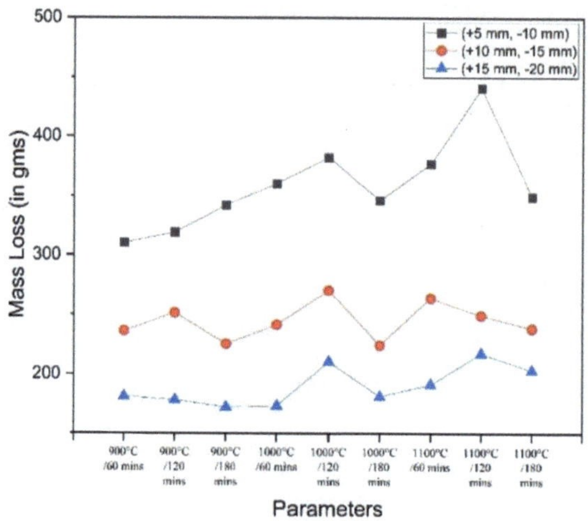

Fig. 4. Weight loss behaviour of manganese ore for different parameters

Plotting the metallisation results based on various ore sizes also allows us to conclude that the metallisation ratio is lower for bigger ore sizes than its substitutes.

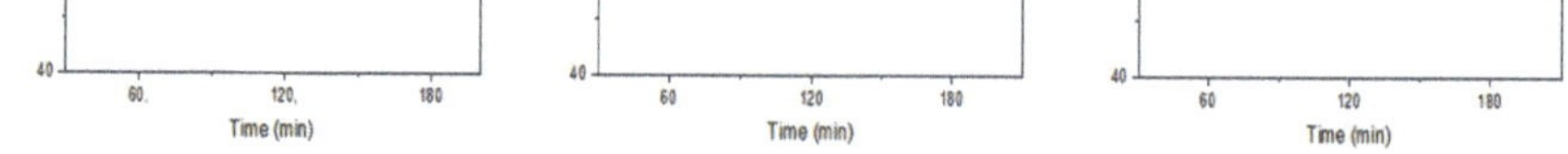

Fig. 5. Metallisation against different parameters for different ore sizes.

3.1 Effect of Temperature

The temperature is a critical factor in the ore reduction process, as it affects the rate and extent of reduction. Generally, higher temperatures lead to faster reduction rates and more complete reduction. However, there is a limit to how high the temperature can be raised before it becomes detrimental to the reduction process. From Fig. 5, it can be observed that with a higher temperature, i.e., greater than 1000 °C, the metallisation decreases.

This supports the literature that chemical reactions control the reaction rate at lower temperatures of up to 1000 °C [13], while diffusion is in charge of the reaction rate at temperatures of 1100 °C and above [14]. Also, at very high temperatures, the reducing agents like coke may react with other components in the system or become oxidised, leading to incomplete reduction or even complete failure of the process. Such conditions are always less preferable for a solid-state reduction system.

High temperatures may also lead to increased energy consumption and processing costs, which is unfavourable on a larger scale model. Therefore, the temperature must be carefully controlled and optimised to ensure efficient and effective reduction.

3.2 Effect of Time

Time can be considered a vital factor in determining the progression of the reduction of ores. From Fig. 5, we can conclude that 120 min provides the best metallisation value for all sizes, which may lead one to believe that a duration longer than 120 min is detrimental to the metallisation of the ores.

Furthermore, longer holding time will lead to higher energy consumption to maintain temperature and push the reaction forward.

4 Conclusion

The following conclusion can be drawn from the reduction-roasting studies on pyrolusite ore:

1. The size of the ore is a very important factor in the reduction process. The results showed that smaller-sized ores are preferably reduced over larger ones under the same reaction conditions.
2. The rise in temperature above 900 °C and the increase in the holding time are not significant factors for the metallization of iron and loss in mass for the fixed size of the ore.
3. The optimum conditions for the maximum Fe metallization are Temperature: 1000 °C, Time: 120 min and Ore size: 5–10 mm.

Thus, the suitable reduction-roasting conditions obtained in our study can be utilized for large-scale tests for energy-efficient smelting operations.

References

1. Chandel, S.S., Singh, P.K., Katiyar, P.K.: NS Randhawa – Mining, Metallurgy & Exploration (2023)
2. Zhang, W., Cheng, C.Y.: Hydrometallurgy **89**, 137–159 (2007)
3. Anacleto, N., Ostrovski, O., Ganguly, S.: ISIJ Int. **44**(9), 1480–1487 (2004)
4. Hofmann, W., Vlajcic, T., Rath, G.: Proceedings Infacon 1989, vol. 1, pp. 185–195. The Ferroalloys Association of the US, New Orleans, LA (1989)
5. Akdogan, G., Hurman, R.: Metall. Mater. Trans. B **26B**, 13–24 (1995)
6. Koursaris, A., See, J.B.: Reactions in the production of high-carbon ferromanganese from Mamatwan manganese ore. National Institute for Metallurgy, Johannesburg, Report no. 1975, 38 p. (1978)
7. Ambade, C., Chandel, S.S.: PK Singh – The International Conference on Metallurgical Engineering and Centenary Celebration (2023)
8. Chandel, S.S., Randhawa, N.S., Singh, P.K.: Mater. Today Proc. (2023)
9. Eric, R.H., Burucu, E.: The mechanism and kinetics of the carbothermic reduction of Mamatwan manganese ore fines. Miner. Eng. **5**(7), 795–815 (1992)
10. Rankin, W.J., van Deventer, J.S.J.: The kinetics of the reduction of manganous oxide by graphite. J. South Afr. Inst. Mining Metall., 239 (1980)
11. Terayama, K., Ikeda, M.: Trans. Jpn Inst. Met. **26**, 108 (1985)
12. Gasik, M.I.: Manganese, Moscow, Metallurgiya (1992)
13. Peretyagin, V.A., Pavlov, A.V.: Metallurgist **47**(9–10), 419–423 (2003)
14. Kor, G.J.W.: Metall. Trans. B **9B**, 307–311 (1978)
15. Olsen, S.E., Olsen, S., Tangstad, M., et al.: Production of Manganese Ferroalloys. Tapir Academic Press, Trondheim, Norway (2007)
16. Gaskell, D.R.: Introduction to Metallurgical Thermodynamics. Scripya Publishing Company, Washington D.C., USA (1940)
17. Karapetyants, M.K.: Examples and Problems in Chemical Thermodynamics. MIR Publishers, Moscow (1974)
18. Samsonov, G.V.: The Oxide Handbook. IFI/PLENUM, USA (1973). Turton, C.N., Turton, T.I. (Trans.)
19. Aytekin, V.: Metallurgy thermodynamics, ITU Metallurgy Faculty Offset Printing Studio, İstanbul (1989)
20. Dikeç, F., Aydın, S.: Problems of solution metallurgy thermodynamics. Offset Atelier of ITU Chemistry and Metallurgy Faculty, Istanbul (1991)
21. Grimsley, W.D., See, J.B., King, P.P.: The mechanism and rate of reduction of Mamatwan manganese ore fines by carbon. J. South Afr. Inst. Min Metall., 51–62 (1977)
22. Mishra, S., Mohapatra, B.K., Dash, N., Rao, D.S.: Ore petrography of low-grade siliceous manganese ore from the Bonai-Keonjhar belt, Orissa, India: the influence of mineral-fabric on their beneficiation. Min. Sci. Technol. **21**, 803–808 (2011)

Iron Recovery Approach from Steel Slag Using Droplet Coalescence Technique

Sheshang Singh Chandel[1,2], Abhinav Maurya[1], Navneet Singh Randhawa[2], and Prince Kumar Singh[1(✉)]

[1] Department of Metallurgical and Material Engineering, Indian Institute of Technology Ropar, Rupnagar 140001, Punjab, India
princeks@iitrpr.ac.in

[2] Metal Extraction and Recycling Division, CSIR-National Metallurgical Laboratory, Jamshedpur 831007, India

Abstract. Currently, the world's utilization rate of steel slag is relatively low, and it is primarily used for landfills, which has a negative impact on the environment and a cost to the economy. Compared to ironmaking slags, steelmaking slag has a relatively high percentage of total iron (between 20 and 35%). The high iron content of slag is a concern for the metalworking industry. Due to the presence of micron-sized iron particles that become trapped, the recovery of iron from slag was limited. Iron entrapment during the reduction of iron oxides in slag systems is a significant issue that reduces the iron yield from the system. This study examines the iron recovery mechanism from molten slag. At the molten stage, the force of attraction between molecules of individual components varies, rendering them an immiscible fluid, particularly metallic iron due to significant difference in density. In this study, iron recovery tracking parameters such as iron agglomeration and its settling time have been examined numerically. Using the commercial CFD software Ansys Fluent, a numerical calculation was performed to predict the size of the metal particle, the settling velocity, and the settling time. A three-dimensional numerical simulation was established using the volume of fluid (VOF) model and slag basicity's ranging from 0.9 to 1.5. In addition, the particle orientation and equivalent droplet radius of metallic iron droplets are examined to support the mechanism and compared with real scenario.

Keyword: Slag Recycling · Computational Fluid Dynamics · Iron Extraction · Droplet Coalescence

Abbreviations

Symbols

$D_{critical}$	Critical Diameter of droplet
F_{drag}	Drag force
g	Gravitational acceleration
m	Molten Iron
p	Pressure
r	Radial length
s	Slag

A. Tiwari et al. (Eds.): iConFIRM 2023, SPPHY 308, pp. 21–27, 2024.
https://doi.org/10.1007/978-981-97-4557-9_3

Greek Symbols

ρ	Density; kg/m^3
μ	Dynamic viscosity; Pa.s
σ	Surface tension; N/m

1 Introduction

Steel is one of the essential materials in human life nowadays. Large consumption of steel has led to proportionate increase in the production of crude steel by industries worldwide. According to past data, every tonne of crude steel leads to 10–15% of steel slag [1], indicating that in 2022, 195.0 MT of steel slag was generated worldwide. Steel slags are rich in minerals like iron (II) oxide (FeO), calcium oxide (CaO), magnesium oxide (MgO), silica (SiO_2), alumina (Al_2O_3), and many more. The presence of high CaO, MgO content makes the steel slag alkaline in nature and mostly used for landfilling. Steel slag used for land filling is the most cost-effective and ecologically benign slag disposal option. However, employing slag for landfilling might gradually damage the dumping ground while also reducing dumping space. Steel slag has another possible application in the cement industry and as rail ballast material, however it is only utilized in trace amounts.

Researchers are continually studying to develop a sustainable technique for recycling steel slag. The primary approach involves the recovery of valuable metals like metallic iron from waste and uses it as a secondary resource to the steelmaking process. Many studies have been conducted to better understand the reduction kinetics of various iron sources. A non-isothermal experiment, known as a thermogravimetric study, was discovered to be one of the most effective methods for studying the reduction kinetics of iron from any material. Iron reduction follows two approaches: using solid carbon, known as direct reduction [2, 3], and gaseous reductants like H_2, CO, etc., known as indirect reduction [4]. From the literature, it was established that direct reduction has a higher reduction efficiency than reduction by gases.

Electric arc furnace (EAF) slag has 25–38% of total iron in the form of FeO, (Fe_2O_3), and $2FeO{\cdot}SiO_2$ and other minor amounts of iron oxide. This research looked on the effective recovery of iron using reduction process on molten slag. The numerical investigation was conducted to study better understand of iron agglomeration and settling time. This study provides a knowledge of how to recover a greater amount of metallic iron, so that it can be used as a secondary iron resource in the steel industry.

2 Modelling and Numerical Parameters

In this investigation, the commercial software ANSYS Fluent 2021 R1 was employed to model the progression and genesis of the iron droplet as iron oxide is reduced from modern slag. The primary objective of the numerical investigation was to gain a comprehensive understanding of the iron agglomeration mechanism and impact of molten iron droplets on settling behavior and its settling time. The computational domain of interest is illustrated in Fig. 1 as a rectangular region with three dimensions, specifically measuring 50 mm × 20 mm × 20 mm. The simulation incorporated one 3 mm droplet and two 3 mm droplets for the purpose of droplet growth. Ansys ICEM software was used to generate the structured mesh with maximum mesh skewness 4.487×10^{-1} and maximum aspect ratio is 1.10. The effect of drop settling and collision behavior was captured using the 100 × 50 × 50 mesh. Table 1 displays the molten iron and slag physical properties. The boundary conditions used in this study are: the computational domain's side and bottom walls are assumed to be slip-free. The upper wall of the computational domain functioned as a pressure outlet. All simulations use atmospheric pressure (Pa = 101325 Pa) and a gravitational force (g_f) of 9.81 m/s along the negative y-axis.

Table 1. Physical properties of reduced molten iron and slag [5].

Material	Density (kg/m^3)	Viscosity (Pa.s)	Surface Tension Coefficient (N/m)	Interface Tension Coefficient Between Molten Iron and Slag (N/m)
Molten Iron	6992	0.0061	1.72	1.24
Slag	2850	0.25	0.54	

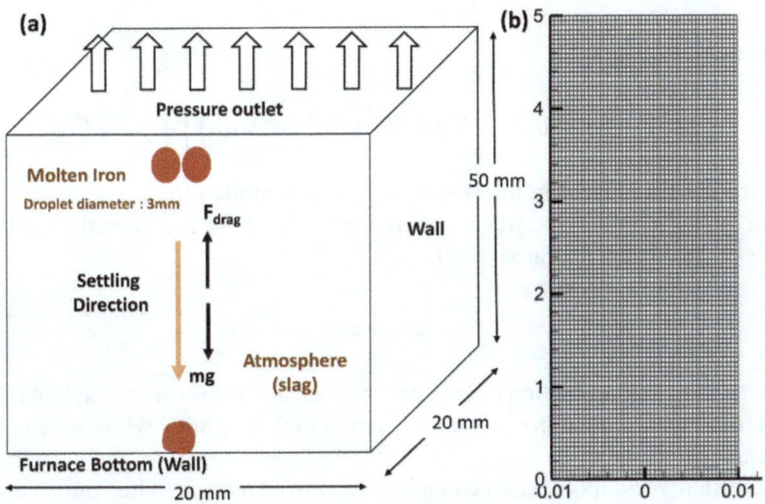

Fig. 1. (a) Schematic diagram of computational domain and (b) structured mesh.

In order to track the interface between the molten iron droplet and the slag, the VOF model is utilized. This is accomplished by solving a single set of momentum equations for both of these water-based fluids. This method monitors the volume fraction of each fluid across the entire computational domain. Consequently, the governing equations are solved in both the domains of the air and molten droplet. There is a range of values for the volume of fraction of fluid in a control volume (F): $F = 0$ indicates that the cell contains only air, $F = 1$ indicates that the cell is full of droplet material, and cells with F values between zero to one ($0 < F < 1$) represent the air-molten droplet interface shown in Eq. 1. The determination of the volume-fraction-averaged continuum density and viscosity in individual control volumes is possible by employing the droplet fluid fraction (F) using Eq. 2 and 3

$$\frac{\partial F}{\partial t} + \nabla(\vec{u}F) = 0 \tag{1}$$

$$\rho = F\rho_d + (1 - F)\rho_{air} \tag{2}$$

$$\mu = F\mu_d + (1 - F)\mu_{air} \tag{3}$$

The equations governing the transport of momentum are subsequently solved by utilising the material properties averaged over the volume fraction. The VOF model is integrated with the momentum conservation equation. The equations governing the conservation of mass and momentum are expressed as follows, under the assumptions of Newtonian flow, incompressible, and laminar behavior.

Continuity Equation [6, 7]

$$\frac{\partial}{\partial t}\rho + \nabla.(\rho\vec{u}) = 0 \tag{4}$$

Momentum Equation [6, 7]

$$\frac{\partial}{\partial t}(\rho\vec{u}) + \nabla.(\rho\vec{u}\vec{u}) = -\nabla\rho + \nabla.\left[\mu\left(\nabla\vec{u} + \nabla\vec{u}^T\right)\right] + \rho\vec{g} + F_{vol} \tag{5}$$

For the two phases present in the system (phase 1 – molten iron, and phase 2 – slag) present in the system, volume fraction of the primary phase is obtained by solving the advection equation as mentioned [6, 7]:

$$\frac{\partial \alpha}{\partial t} + u.\nabla\alpha = 0 \tag{6}$$

α represents the volume fraction of iron phase in each control volume. When α is equal to 1, it is filled with molten iron; when α is equal to 0, it is filled with molten slag, and u represents the velocity.

The continuity and momentum equations are solved using a finite volume approach. The QUICK scheme was used to minimize numerical diffusion while solving the momentum equation. The pressure implicit scheme was used to combine pressure and velocity. The transient model, with a time step of 0.0001 s and relaxation factors of 0.3 and 0.7, was used for pressure and momentum convergence. The scaler residual of 10^{-6} is used as the convergence criterion for all governing equations. Each simulation case was run for 1 s.

3 Result and Discussion

Impact of iron agglomeration was mainly examined to understand the mechanism of particle growth and effect on settling time. The two droplets of 3 mm were considered to examine the stages of iron growth during the reduction process. Figure 2 shows the phase contour of molten iron droplet during its agglomeration, which takes place in four major stages: Approaching, Film formation, coalescence, and Relaxation. In the first stage, the particles at critical distance approach towards each other. Once the droplets are close, surrounding fluid between the droplets forms a film. Due to formation of film, stress is developed on the spherical droplet surface which leads to change in droplet shape locally. On proceeding further, the film thickness reduces and a thin bridge forms between the droplets as the fluid present between droplets is displaced due to the pressure difference. Once the drop coalescence is complete, shape of droplet fluctuates due to the surface tension by the surrounding fluid. When the effect of surface tension becomes nominal, the fluctuation and vibration of the drop achieves equilibrium. Figure 2 (a) and (b) show the phase contour and vector of a drop collision.

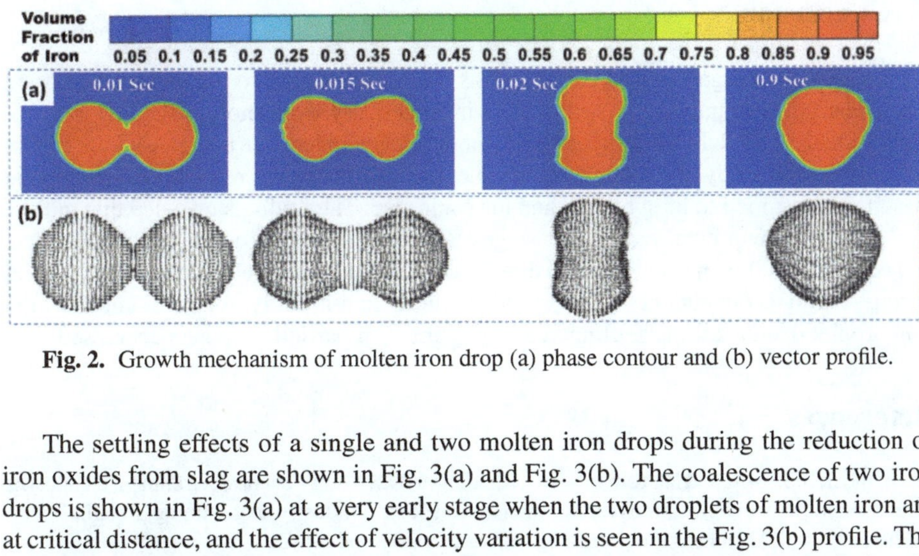

Fig. 2. Growth mechanism of molten iron drop (a) phase contour and (b) vector profile.

The settling effects of a single and two molten iron drops during the reduction of iron oxides from slag are shown in Fig. 3(a) and Fig. 3(b). The coalescence of two iron drops is shown in Fig. 3(a) at a very early stage when the two droplets of molten iron are at critical distance, and the effect of velocity variation is seen in the Fig. 3(b) profile. The turbulence seen during the collision of two droplets was also captured in numerical study and shown in Fig. 3(b). The effect of droplet settling can also be calculated using Stoke's equations. $V_1 = \frac{gd^2(\rho_m - \rho_s)}{18\mu_s}$, which is valid only for spherical droplet. For non-spherical droplet the Stoke's equation was modified to $V_2 = \frac{2gd^2(\rho_m - \rho_s)}{3\mu_s} \times \frac{\mu_s + \mu_m}{\mu_s + 3\mu_m}$ [8]. As per the Stoke's and modified Stoke's equations, the velocity was calculated for 2 mm droplet as ~0.5 m/s which increases up to ~0.6 m/s after drop coalescence.

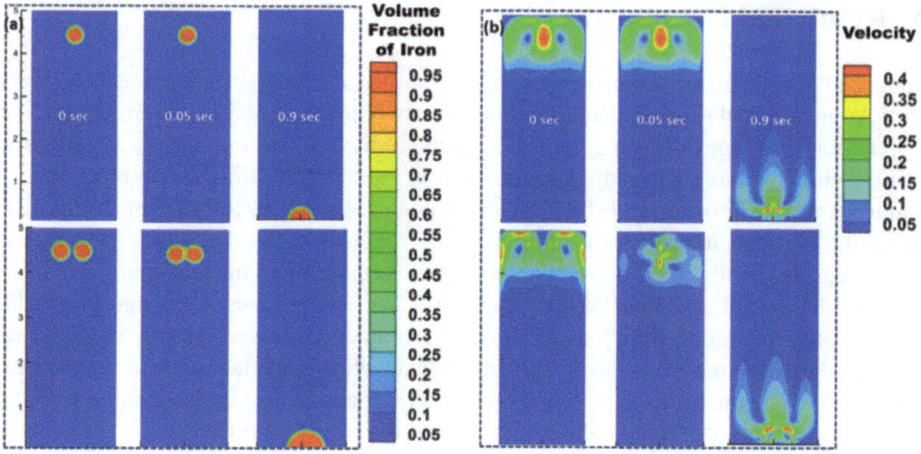

Fig. 3. (a) Single and Two droplet settling contour and (b) its velocity contour.

4 Conclusion

A numerical investigation was conducted as part of this study to better understand the mechanism of iron particle growth during iron recovery from molten slag. This three-dimensional study with volume-of-fluid model was carried out using ANSYS Fluent 2021 R1. In this study, the molten iron droplet present during the reduction of iron oxide in slag was assumed to be spherical and the same size. The findings revealed that molten iron droplet growth occurred in four stages: approaching, film formation, coalescence, and equilibrium. It can also be seen using stoke equations, and the velocity of the droplet during settling is dependent on the droplet diameter in this study. As the diameter of the iron droplet decreases, the settling velocity of the iron particle also gets decreased.

References

1. Chandel, S.S., Singh, P.K., Katiyar, P.K., Randhawa, N.S.: A review on environmental concerns and technological innovations for the valorization of steel industry slag. Mining Metall. Explor. **2023**, 1–28 (2023). https://doi.org/10.1007/S42461-023-00886-Z
2. Singh Chandel, S., Singh Randhawa, N., Kumar Singh, P.: Thermodynamic and kinetic aspect of solid state reduction of Electric Arc Furnace slag through coke: an experimental study. Mater. Today Proc. (2023). https://doi.org/10.1016/J.MATPR.2023.07.209
3. Ambade, C., Chandel, S.S., Singh, P.K.: Reduction kinetics of composite steel slag-coke pellets. In: The International Conference on Metallurgical Engineering and Centenary Celebration, pp. 52–57 (2024). https://doi.org/10.1007/978-981-99-6863-3_7
4. Monazam, E.R., Breault, R.W., Siriwardane, R.: Reduction of hematite (Fe2O3) to wüstite (FeO) by carbon monoxide (CO) for chemical looping combustion. Chem. Eng. J. **242**, 204–210 (2014). https://doi.org/10.1016/j.cej.2013.12.040
5. Ohashi, M., Maruishi, T., Toramaru, A.: Coalescence of growing bubbles in highly viscous liquids. Geochem. Geophys Geosyst. **23**(11) (2022). https://doi.org/10.1029/2022GC010618

6. Chandel, S.S., Maurya, A., Singh, P.K.: Numerical investigation of quenching technique for steel alloy hardening process using twins liquid jets. Mater. Today Proc. **62**(P14), 7348–7352 (2022). https://doi.org/10.1016/J.MATPR.2022.01.368
7. Paulraj, M.P., Chandel, S.S., Kumar, A., Sahu, S.K.: Numerical investigation to enhance the thermal performance of solar air dryers using parallelogram type transverse ribs. Trans. Indian Natl. Acad. Eng. **2023**, 1–16 (2023). https://doi.org/10.1007/S41403-023-00431-W
8. Zhang, H., Li, B., Wei, Y., Wang, H.: The settling behavior of matte particles in copper slag and the new technology of copper slag cleaning. J. Mater. Res. Technol. **15**, 6216–6230 (2021). https://doi.org/10.1016/j.jmrt.2021.11.061

Copper Slag as a Source of Iron: An Overview

Jayati Priya[1,2], Navneet Singh Randhawa[2], and Manish Kumar Jain[1](✉)

[1] Department of Environmental Science and Engineering, IIT(ISM) Dhanbad, Dhanbad 826004, India
manish@iitism.ac.in

[2] Metal Extraction and Recycling Division, CSIR-National Metallurgical Laboratory, Jamshedpur 831007, India

Abstract. The production of Copper slag occurs during the smelting of Cu, and it contains various compounds, including oxides of Fe, Cu, Co, Al, Si, Ni, Cr, and Ca. A significantly smaller percentage is utilised for low-end applications and is mainly dumped, thus occupying large spaces for a longer duration. As a result, gainful utilization of such slag into value-added products is necessary to avoid the associated environmental hazards and economic loss. Notably, Cu slag possesses more valuable metals like Cu and Fe than certain raw ores. This paper explores methods for recovering and treating copper slag. Metal recovery has been a focal point in recent decades but comes with challenges. The process consumes significant energy, and its low recovery ratio hampers the advancement of recovery technologies, making them economically impractical and difficult to implement on an industrial scale. The high temperature of copper slag derived from furnaces (>1300 °C) indicates the potential for reclaiming substantial energy. Integrating waste heat and metal recovery techniques makes it possible to address the issues above. However, most research on recovery lacks an economic analysis of their technologies. Due to its high Fe content, copper slag can be a Fe-adjusting material in cement clinker production. However, before utilizing Cu slag as a material resource, it is crucial to perform simultaneous metal extraction.

Keywords: Copper Slag · Recycled material · Circular economy · Review · Reduction

1 Introduction

Waste management is a crucial and one of the major concerns of the metallurgical industries, and one such waste material is copper slag. Copper slag is a by-product of copper manufacturing that is obtained through a process called pyrometallurgy. For each ton of copper metal produced worldwide, about 2.2–3.0 tons of slag is produced, estimated to be approximately 40 million tons of slag generation annually (J. P. Wang & Erdenebold, 2020), with the content of iron reaching 45%; the copper slag has potential to produce approximately 26.8 million tons of Fe annually (Heo et al., 2013). A significantly smaller percentage is utilized for low-end applications and is mainly dumped, thus occupying large spaces for a longer duration. As a result, gainful utilization of such slag

A. Tiwari et al. (Eds.): iConFIRM 2023, SPPHY 308, pp. 28–38, 2024.
https://doi.org/10.1007/978-981-97-4557-9_4

into value-added products has become a challenging task for the industry. In the past few decades, numerous attempts have been made by researchers for the reuse and recycling of this slag through different processes and products (Chandel et al., 2023).

Several studies have explored the utilization of copper slag as a source of iron, highlighting its potential for iron recovery and recycling. They have created a new technology that improves the process of extracting copper and iron components from copper slag, emphasizing the promotion of mineralization of valuable minerals and the induction of mineral grain growth (Guo et al., 2016). Similarly, discussed the high iron content of copper smelting slag and its potential applications in value-added products, such as abrasive tools, roofing granules, and road-base construction, emphasizing the potential for recycling metal values from the slag (Wang & Erdenebold, 2020). Furthermore, highlighted the significant iron content in copper slag, emphasizing the possibility of recovering and recycling iron along with copper. It recovers copper as an iron-copper alloy and oxides in the final slag (Busolic et al., 2011). Moreover, studies have proposed various methods for iron recovery from copper slag. For instance, proposed the recovery of value metals from copper slag through ammonium chloride treatment, addressing the challenge of extracting iron from the slag using traditional processes (Zhou et al., 2015). Additionally, demonstrated the potential of copper slag as a source of valuable metals, including iron, copper, and zinc, and proposed a method for separating and recycling these metals from dumped copper slag using microwave irradiation (Liao et al., 2016). These studies collectively underscore the potential of copper slag as a secondary source of iron and the various methods for its recovery and utilization.

Furthermore, the environmental and socioeconomic impact of utilizing copper slag as an alternative source of iron has been discussed. Emphasized the environmental benefits of using copper slag as a source of iron, highlighting its potential to revamp and rejuvenate the environment where the slag was previously dumped and stored (Gabasiane et al., 2021). Industrial practices are increasingly shifting towards sustainable and eco-friendly methods, which is aligned with the growing emphasis on environmental responsibility. In conclusion, the literature review indicates a growing interest in utilizing copper slag as a potential source of iron. The studies reviewed highlight the high iron content of copper slag and various methods for its recovery and utilization, emphasizing its potential for iron recycling and its environmental benefits.

1.1 Properties

The copper slag particles are blackish and possess a shiny surface. The table provided outlines the typical physical attributes of standard copper slag (Table 1).

The physical characteristics of copper slags are significantly influenced by the method of cooling applied. Cooling molten copper slag by immersing it in water creates amorphous granulated slag upon discharge from smelting facilities. This granulated slag is typically composed of small, well-defined, and angular particles, with sizes ranging from 4.75 mm to 0.075 mm (Gorai et al., 2003). The porous nature of granulated slags contributes to their lower weight. In contrast, slags cooled in open air are characterized by a coarse, dense, compact, and crystalline structure (Gabasiane et al., 2021).

Table 1. Physical properties of typical copper slag (Kundu et al., 2023)

Properties	Appearance	Bulk Density	Moisture	Density	Internal friction angle	Mohr's scale of hardness
Value	Black, glossy	2.3–2.6 (T/m^3)	<1%	2.8–3.8 (g/cm^3)	40–53	6–7

Copper slags are primarily linked to harmful and noxious metals, including Cu, Pb, Mo, Zn, Co, etc. (Sarfo et al., 2017). These undesirable components render copper slag inappropriate for disposal in landfills. Disposing of such slags in this manner would pose environmental risks, contaminating land and groundwater. Given these circumstances, it is imperative to prioritize the retrieval, recycling, and utilization of these waste products. Existing literature documents instances where copper slags have been used to recycle Fe using different extraction mechanisms, including Hydro-metallurgy and pyro-metallurgy.

1.2 Background and Context

The copper slag produced by the smelting industry contains valuable metals like Cu, Fe, Zn, Pb, Co, and Ni, along with harmful components like As, Hg, Bi, and Sb. However, over 80% of global copper slag is discarded without proper disposal, wasting valuable metals and harming the environment (Phiri et al., 2022). While some copper slag has been used in civil engineering applications, its high metal content limits its use. Various processes, such as flotation, leaching, reduction roasting, magnetic separation, smelting reduction, and molten modification, have been identified for recycling valuable metals from different copper slags (Guo et al., 2016). Addressing technical and environmental challenges is crucial for promoting sustainable advancement in the treatment procedures for copper slag, driving industrial progress, and improving metallurgical recycling. Researchers have explored strategies to address the challenge of separating Iron and copper, which tend to amalgamate during reduction. Some have employed flotation processes to recover copper-bearing phases before or after producing pig iron (Guo et al., 2016). Alternatively, treatment with aluminum sulfide-ferrous sulfide flux has been investigated to obtain a matte phase, aiding in copper removal from pig iron. This prepares the Iron for subsequent use in steelmaking, partially overcoming the energy-intensive separation of Iron and copper (Ortega-López et al., 2014). Utilizing iron-copper alloy, obtained through reduction, offers a valuable alternative to costly electrolytic copper in weather-resistant steel production (Xian-Lin et al., 2015).

In the reduction treatment for copper slag, carbonaceous materials are employed alongside additives for promotion in carbothermic reduction roasting or smelting processes (Łabaj et al., 2021). Biomass, such as pine sawdust, corncob, straw, and walnut shells, emerges as a renewable energy source with potential advantages over conventional coal (Zuo et al., 2016). This includes reducing dependence on traditional coals and lowering emissions. During the metallurgical recycling and cleaner treatment process for copper slag, removing and immobilizing harmful elements is essential to obtain a cleaned secondary slag with low residual toxic elements. Hydrometallurgical processes are commonly used due to their high extraction efficiency for valuable metals and harmful elements. To separate, remove, or recycle elements, they are first dissolved in either an acid or alkali solution. Then, they are filtered, washed, and extracted using specific extraction agents (Mikula et al., 2021). The amount of certain elements present in the secondary slag is significantly reduced as a result of this process. Though, it is essential to consider the large portion of liquid waste in the form of acidity or alkalinity generated during the hydrometallurgical processes to avoid secondary pollution of the environment.

Unlike hydrometallurgical procedures, pyrometallurgical methods often exhibit distinct tendencies in eliminating environmentally harmful trace elements. The reduction stages involved the utilization of waste ferric-oxide desulphurize (W.F.D.) with components comprising 17.91% CaO, 5.82% C, 32.33% S and 7.89% Fe during the smelting reduction-sulphurising process aimed at effectively purifying waste copper slag (Guo et al. et al., 2018). The incorporation of W.F.D., ranging from 0 to 15%, significantly enhanced the extraction rates of Sb, Ni, and As, increasing the percentages from 52.23% to 95.51%, 43.21% to 98.94%, and 45.56% to 96.81%, correspondingly. This improvement was attributed to transforming Sb, Ni, and As minerals within the copper slag into sulfide or metal forms through reduction and sulfidation processes, followed by their subsequent migration and settling within the matte phase. Furthermore, the influence of W.F.D. on the elimination of Pb, Zn, Hg, and Bi was moderate, resulting in a removal rate exceeding 90%. This was primarily due to converting Pb, Zn, Pb, and Bi minerals into vaporous states during the smelting reduction, driven by their high saturated vapor pressures.

As a result, these elements can be effortlessly reclaimed from the cooled exhaust gas, rendering their recycling feasible. The toxicity characteristic leaching procedure (TCLP) test was used to evaluate the environmental impact of the purified residual slag, demonstrating that the detrimental attributes of the leachate remained within the established thresholds of USEPA-2017. This validation confirmed the safety of the decontaminated secondary slag containing residual harmful constituents for human and environmental well-being (Table 2).

Table 2. Existing literature on recovery of iron from copper slag

Author(s)	Objectives	Methods Used	Key Findings
(Busolic et al., 2011)	A two-step process to recover an iron-rich alloy	Smelting	This study uses coke as a reductant to reduce Flash Smelting slag, surpassing the carbon requirement by up to 150%. The results demonstrate a two-step process that recovers an iron-rich alloy with over 98% iron and copper ranging from 0.6% to 0.8%

(*continued*)

Table 2. *(continued)*

Author(s)	Objectives	Methods Used	Key Findings
(Erdenebold et al., 2018)	An investigation aimed to extract pig iron from copper smelting slag through a reduction smelting technique	Smelting	At temperatures below 1400 °C, some portions of the copper smelting slag agglomerated. Pig iron recovery increased with higher smelting temperature and duration, varying significantly depending on the reaction temperature
(L. Wang et al., 2023)	This study aimed to increase metal recovery efficiency from Copper Slag (CS) by gaining insights into its primary elemental phases	Leaching	pH-dependent leaching experiments revealed that undissolved Fe and Zn are present in fayalite, magnetite, calcium-iron silicates, and sphalerite. Mineral liberation analysis showed that most Fe and Zn are contained in fayalite. Future efforts for the recovery of Zn and Fe should not solely focus on fayalite, but also consider the impact of calcium-iron silicate and iron oxide
(Li et al., 2021)	Extract valuable elements (copper, iron, lead, and zinc) from waste copper slag	Flotation, pelletization, reduction, and magnetic separation	This study presents a systematic approach to The process that involves flotation concentration to obtain a copper concentrate with 21.50% Cu and 77.78% recovery, followed by pelletization, reduction, and magnetic separation to obtain a magnetic concentrate with 90.21% total iron and 0.4% Cu, and a recovery rate of 91.34% for iron and 83.41% for copper. The non-magnetic tailings are mixed with clinker and standard sand to produce ordinary Portland cement. The process yields several products, including flotation concentrate, magnetic concentrate, direct reduction dust, and ordinary Portland cement

(continued)

Table 2. (*continued*)

Author(s)	Objectives	Methods Used	Key Findings
(Lykasov et al., 2016)	In extracting copper from sulfide ore, residue known as slag is generated at the processing plant	Reducing a layered batch of slag with 5 mm thickness layers interspersed with coal interlayers	This study analyzes slag phase changes when exposed to carbon gasification byproducts at 1100 °C and 1200 °C . The results show that carbon gasification at 1100 °C converts iron in the slag into metallic form, with further improvement at 1200 °C . Indirect reduction above the melting point leads to metallic iron accumulation on the outer surface, forming sizable inclusions that can be easily separated with magnetic processes
(Kim et al., 2013)	A process of separating iron from waste copper slag using physicochemical methods for upgrading purposes	Reduction followed by magnetic separation	The process involves crushing, reduction reaction at 1250 °C , and dry magnetic separation to yield a magnetic product with over 65% iron under 0.03 T magnetic field strength, with an iron recovery exceeding 85%
(Zhang et al., 2020a)	Recovery of Iron from Copper Slag	Coal-Based Direct Reduction	Copper slag was directly reduced by coal, producing metallic iron. Fe3O4 and Fe2SiO4 were reduced, achieving a 91.55% Fe grade and 98.13% recovery. Reduction was controlled by interfacial reaction and carbon gasification at 1050 °C , and gas diffusion at higher temperatures. Activation energy increased as reduction proceeded, with Fe2SiO4 requiring more energy than Fe3O4
(Erdenebold and Wang, 2020)	Recycling into Metal Values	Reduction of Copper Smelting Slag by Carbon	This study recovered Fe-Cu alloy and slag for cement from copper slag. When carbon is used at temperatures higher than 1400 °C, pig iron containing copper can be obtained through reduction smelting. The addition of CaO reduces iron oxide and changes the slag's composition. When pig iron and copper oxide are mixed, the copper oxide dissolves and the zinc oxide turns into a gas and evaporates. When CaO is added above 5wt.%, acid slag turns into calcium silicate slag similar to blast furnace slag

(*continued*)

Table 2. *(continued)*

Author(s)	Objectives	Methods Used	Key Findings
(Sarfo et al., 2017)	To utilize the decontaminated copper in other domains effectively	Reduction process	In order to utilize the decontaminated copper in other domains effectively, we must consider chemical and physical attributes. When we leach secondary copper slag with reducing agents; the metal content specially Iron decreased drastically as compared to the initial slag. On the other hand, the content of silica, calcium, and aluminum increased. The secondary slag shows the superior performance because of its hardness (>657.4 VHN) and surface density (<2.73 g/cm^3); it can be used in glass, proppant, and ceramic industries
(Cao et al., 2016)	Iron Recovery from Discarded Copper Slag	Direct reduction, followed by grinding and magnetic separation	This study shows how to reduce carbon-bearing pellets from discarded copper slag. The best conditions for the process are a mass proportion of 100:25:10:3 for slag, reduction coal, limestone, and industrial sodium carbonate. To reduce the pellets, the furnace should be heated to a temperature of 1280 °C. This process took 35 min and three layers of pellets is used. The iron products obtained had an iron grade of 90.35% with an iron recovery rate of 89.70%. The process of reducing copper slag into metallic iron involves the reduction of fayalite and magnetite.

2 Economical Aspects

The recovery of iron from copper slag is an economically feasible and environmentally significant process due to the increasing volume of slag in primary copper production and the potential for metal losses in discarded slag. Various studies have explored the feasibility of recovering iron from copper slag using different methods such as (bio)hydrometallurgical processes, bio-flotation, bio-leaching, and concentrated solar power (CSP) (Potysz and Kierczak, 2019; Behera et al., 2022; González et al., 2021; Avarmaa et al., 2019). These studies emphasize the importance of developing environment-friendly, techno-economic, and efficient processes to utilize abandoned copper slag wastes for metals especially iron recovery (Potysz & Kierczak, 2019). The recovery of valuable metals from copper slag, including copper, gold, platinum, palladium, iron, and magnetite, has been demonstrated through techniques such as froth flotation, bio-flotation, and slow cooling, with promising results (Busolic et al., 2011; Avarmaa et al., 2019; Fan et al., 2014; Gao et al., 2020; Štirbanović & Marković, 2011). Additionally, copper, cobalt, and zinc recovery from copper smelter slag using oxidizing leaching and solvent extraction have shown high metal recovery percentages (Hughes, 2000).

Several studies have highlighted the economic feasibility of recovering metal from copper slag. A preliminary economic analysis has shown that copper recovery alone could generate significant economic benefits. This emphasizes the potential profitability of metal recovery from copper slag (González et al., 2021). Using smelting technology to recover metals from slag is economically justified if the value of recovered metals exceeds a certain threshold per ton of slag treated (Mitrašinović & Wolf, 2015). Various studies have shown that recovering metals from copper slag is economically feasible, with high iron recovery percentages and potential economic benefits. It is important to develop efficient and environmentally friendly metal recovery processes from copper slag to ensure resource efficiency and environmental protection.

3 Conclusion

Copper slag, viewed as waste, has potential for practical use. This study aims to identify approaches for mitigating copper slag's environmental hazards through recycling or reuse. Pyrometallurgy is crucial for metal production, but it harms the environment. Slag disposal can cause metal mobilization and weathering, negatively impacting the environment. Recovering valuable metals from slag is a promising alternative to reduce the required surface areas and minimize environmental concerns. Using these materials as secondary resources benefits both the industry and the environment.

Acknowledgements. The authors are thankful to the Director, IIT(ISM), Dhanbad, for the necessary permission and the organizing committee of iConFIRM-2023 for allowing us to present this paper.

References

Avarmaa, K., Klemettinen, L., O'Brien, H.: Urban mining of precious metals via oxidizing copper smelting. Miner. Eng., **133**, 95–102 (2019). https://doi.org/10.1016/j.mineng.2019.01.006

Avarmaa, K., O'Brien, H., Klemettinen, L., Taskinen, P.: Precious metal recoveries in secondary copper smelting with high-alumina slags. J. Mater. Cycles Waste Manage. **22**(3), 642–655 (2019). https://doi.org/10.1007/s10163-019-00955-w

Behera, S., Panda, S., Mulaba-Bafubiandi, A.: Valorization of copper smelter slag through the recovery of metal values by a synergistic bioprocess system of bio-flotation and bio-leaching. Environ. Qual. Manage. **32**(1), 233–241 (2022). https://doi.org/10.1002/tqem.21885

Busolic, D., Parada, F., Parra, R., Sanchez, M., Palacios, J., Hino, M.: Recovery of iron from copper flash smelting slags. Trans. Inst. Mining Metall. Sect. C Miner. Process. Extr. Metall. **120**(1), 32–36 (2011). https://doi.org/10.1179/037195510X12772935654945

Cao, Z., Sun, T., Xue, X., Zhan-hua, L.: Iron recovery from discarded copper slag in a RHF direct reduction and subsequent grinding/magnetic separation process. Minerals **6**(4), 119 (2016). https://doi.org/10.3390/min6040119

Chandel, S.S., Singh, P.K., Katiyar, P.K., et al.: A review on environmental concerns and technological innovations for the valorization of steel industry slag. Mining Metall. Exp. **40**, 2059–2086 (2023). https://doi.org/10.1007/s42461-023-00886-z

Das, B., Mishra, B., Angadi, S., Pradhan, S., Prakash, S., Mohanty, J.: Characterization and recovery of copper values from discarded slag. Waste Manag. Res. J. Sustain. Circular Econ. **28**(6), 561–567 (2009). https://doi.org/10.1177/0734242x09343943

Erdenebold, U., Wang, J.: A study on reduction of copper smelting slag by carbon for recycling into metal values and cement raw material (2020). https://doi.org/10.20944/preprints202001.0177.v1

Erdenebold, U., Choi, H.M., Wang, J.P.: Recovery of pig iron from copper smelting slag by reduction smelting. Arch. Metall. Mater. **63**(4), 1793–1798 (2018). https://doi.org/10.24425/amm.2018.125106

Fan, Y., Shibata, E., Iizuka, A., Nakamura, T.: Crystallization behaviors of copper smelter slag studied using time-temperature-transformation diagram. Mater. Trans. **55**(6), 958–963 (2014). https://doi.org/10.2320/matertrans.m-m2014819

González, D., Prazuch, J., Ruiz-Bustinza, I., González-Gasca, C., Gómez-Rodríguez, C., González, L.: Recovery of copper and magnetite from copper slag using concentrated solar power (CSP). Metals **11**(7), 1032 (2021). https://doi.org/10.3390/met11071032

Gabasiane, T.S., Danha, G., Mamvura, T.A., Mashifana, T., Dzinomwa, G.: Environmental and socioeconomic impact of copper slag—a review. Crystals **11**(12) (2021). https://doi.org/10.3390/cryst11121504

Gorai, B., Jana, R.K., Premchand: Characteristics and utilisation of copper slag - a review. Resour. Conserv. Recycl. **39**(4), 299–313 (2003). https://doi.org/10.1016/S0921-3449(02)00171-4

Guo, Z., Zhu, D., Pan, J., Wu, T., Zhang, F.: Improving beneficiation of copper and iron from copper slag by modifying the molten copper slag. Metals **6**(4) (2016). https://doi.org/10.3390/met6040086

Guo, Z., Zhu, D., Pan, J., Zhang, F., Yang, C.: Industrial tests to modify molten copper slag for improvement of copper recovery. JOM **70**(4), 533–538 (2018). https://doi.org/10.1007/s11837-017-2671-5

Gao, X., Chen, Z., Shi, J., Taskinen, P., Jokilaakso, A.: Effect of cooling rate and slag modification on the copper matte in smelting slag. Mining Metall. Exp. **37**(5), 1593–1601 (2020). https://doi.org/10.1007/s42461-020-00274-x

Heo, J.H., Kim, B.S., Park, J.H.: Effect of CaO addition on iron recovery from copper smelting slags by solid carbon. MMTB **44**, 1352–1363 (2013)

Hughes, S.: Applying ausmelt technology to recover Cu, Ni and Co from slags. J. Met. **52**(8), 30–33 (2000)

Kim, B.S., Jo, S.K., Shin, D., Lee, J.C., Jeong, S.B.: A physico-chemical separation process for upgrading iron from waste copper slag. Int. J. Miner. Process. **124**, 124–127 (2013). https://doi.org/10.1016/j.minpro.2013.05.009

Kundu, T., Senapati, S., Das, S.K., Angadi, S.I., Rath, S.S.: A comprehensive review on the recovery of copper values from copper slag. Powder Technol. **426**. (2023). https://doi.org/10.1016/j.powtec.2023.118693

Klemettinen, L., Aromaa, R., Dańczak, A., O'Brien, H., Taskinen, P., Jokilaakso, A.: Distribution kinetics of rare earth elements in copper smelting. Sustainability **12**(1), 208 (2019). https://doi.org/10.3390/su12010208

Łabaj, J., Blacha, L., Jodkowski, M., Smalcerz, A., Fröhlichová, M., Findorak, R.: The use of waste, fine-grained carbonaceous material in the process of copper slag reduction. J. Cleaner Prod. **288** (2021). https://doi.org/10.1016/j.jclepro.2020.125640

Li, K., Ping, S., Wang, H., Wen, N.: Recovery of iron from copper slag by deep reduction and magnetic beneficiation. Int. J. Miner. Metall. Mater. **20**(11), 1035–1041 (2013). https://doi.org/10.1007/s12613-013-0831-3

Li, S., Guo, Z., Pan, J., Zhu, D., Dong, T., Lu, S.: Stepwise utilization process to recover valuable components from copper slag. Minerals **11**(2), 1–14 (2021). https://doi.org/10.3390/min11020211

Liao, Y., Zhou, J., Huang, F.: Separating and recycling of Fe, Cu, Zn from dumped copper slag by microwave irradiation assisted carbothermic method. J. Residuals Sci. Technol. **13**, 155–160 (2016)

Lykasov, A.A., Ryss, G.M., Sharafutdinov, D.G., Pogodin, A.Y.: Extraction of iron from copper-plant slag. Steel Trans. **46**(9), 609–613 (2016). https://doi.org/10.3103/S0967091216090059

Mitrašinović, A., Wolf, A.: Separation and recovery of valuable metals from nickel slags disposed in landfills. Sep. Sci. Technol. 150615133933004 (2015). https://doi.org/10.1080/01496395.2015.1056360

Mikula, K., Izydorczyk, G., Skrzypczak, D., Moustakas, K., Witek-Krowiak, A., Chojnacka, K.: Value-added strategies for the sustainable handling, disposal, or value-added use of copper smelter and refinery wastes. J. Hazard. Mater. **403**, 123602 (2021). https://doi.org/10.1016/j.jhazmat.2020.123602

Ortega-López, V., Manso, J.M., Cuesta, I.I., González, J.J..: The long-term accelerated expansion of various ladle-furnace basic slags and their soil-stabilization applications. Constr. Build. Mater. **68**, 455–464 (2014). (October). Elsevier Ltd

Potysz, A., Kierczak, J.: Prospective (bio)leaching of historical copper slags as an alternative to their disposal. Minerals **9**(9), 542 (2019). https://doi.org/10.3390/min9090542

Phiri, T.C., Singh, P., Nikoloski, A.N.: The potential for copper slag waste as a resource for a circular economy: a review – Part I. Miner. Eng. **180** (2022). https://doi.org/10.1016/j.mineng.2022.107474

Sarfo, P., Das, A., Wyss, G., Young, C.: Recovery of metal values from copper slag and reuse of residual secondary slag. Waste Manage. **70**, 272–281 (2017). https://doi.org/10.1016/j.wasman.2017.09.024

Stirbanović, Z., Marković, Z.: The effect of copper bearing particles liberation on copper recovery from smelter slag by flotation. Sep. Sci. Technol. **46**(16), 2496–2500 (2011). https://doi.org/10.1080/01496395.2011.599355

Tang, C., Li, K., Wen, N., Fan, D.: Recovering iron from iron ore tailings and preparing concrete composite admixtures. Minerals **9**(4), 232 (2019). https://doi.org/10.3390/min9040232

Wang, J.P., Erdenebold, U.: A study on reduction of copper smelting slag by carbon for recycling into metal values and cement raw material. Sustainability (Switzerland) **12**(4) (2020). https://doi.org/10.3390/su12041421

Wang, L., Su, Q., Tian, H., Du, Y.: Leaching behavior and occurrence of metal elements in copper slag: the key to recycling metals in copper slag. J. Hazard. Mater. Adv., 100374 (2023). https://doi.org/10.1016/j.hazadv.2023.100374

Zhou, X., Zhu, D., Pan, J., Tang, W.: Utilization of waste copper slag to produce directly reduced iron for weathering resistant steel. ISIJ Int. **55**(7), 1347–1352 (2015). https://doi.org/10.2355/isijinternational.55.1347

Zuo, Z., et al.: Thermogravimetric study of the reduction of copper slag by biomass: reduction characteristics and kinetics. J. Therm. Anal. Calorim. Calorim. **126**(2), 481–491 (2016). https://doi.org/10.1007/s10973-016-5570-z

Zhang, G., Wang, N., Chen, M., Cheng, Y.: Recycling nickel slag by aluminum dross: iron-extraction and secondary slag stabilization. ISIJ Int. **60**(3), 602–609 (2020). https://doi.org/10.2355/isijinternational.isijint-2019-173

Zhang, H., Hu, C., Gao, W., Lu, M.: Recovery of iron from copper slag using coal-based direct reduction: reduction characteristics and kinetics. Minerals **10**(11), 973 (2020). https://doi.org/10.3390/min10110973

Machine Learning Approach for Accurate Slag Eye Predictions in Steelmaking Ladles

Somenath Mukherjee and Vishnu Teja Mantripragada(✉)

Department of Fuel Minerals and Metallurgical Engineering, Indian Institute of Technology (ISM) Dhanbad, Jharkhand 826004, India
vishnuteja@iitism.ac.in

Abstract. In the arena of steelmaking process optimization, managing slag-eye formation in gas-stirred ladles is critical. The slag eye area in steelmaking ladles is a critical parameter influencing the efficiency of inclusion removal and the control of slag-metal interactions. This study introduces a machine learning methodology for predicting the slag eye area, which has traditionally been a challenge due to the complex, dynamic nature of the ladle metallurgical process. Various machine learning models were evaluated using a comprehensive dataset from the available literature derived from physical and mathematical modeling. The model incorporates parameters such as slag thickness, melt height, ladle dimensions, number and position of gas inlets, physical properties of the constituent phases, etc., to estimate the slag eye area in both single and dual-purged steelmaking ladles. This research demonstrates superior slag eye area predictive capabilities, with implications for enhancing energy efficiency, alloy recovery, and overall process efficiency in the steelmaking industry.

Keywords: Steelmaking Ladle · Slag Eye · Machine Learning

1 Introduction

In steelmaking ladles, argon gas is introduced for homogenizing the bath composition and temperature, alloy mixing, inclusion removal, etc. The formation of the slag eye occurs as argon bubbles ascend due to buoyancy, moving from the molten steel through the overlying slag layer to the surface. The presence of slag eye leads to increased heat loss from the exposed molten steel surface, which can result in temperature drops and reduced energy efficiency. Additionally, slag eye exposes the molten steel to the atmosphere, increasing the risk of re-oxidation and nitrogen pickup, which can deteriorate steel quality. Therefore, while argon purging is a crucial step in refining steel and ensuring its homogeneity and cleanliness, controlling the formation and size of the slag eye is equally important to minimize negative impacts on the steelmaking process and the final product quality.

The size of the slag eye affects various aspects of steelmaking, including heat loss, alloying, and the removal of inclusions [1, 2]. Conducting plant trials directly in operational steel plants offers real-time data and insights into slag behavior under actual

A. Tiwari et al. (Eds.): iConFIRM 2023, SPPHY 308, pp. 39–46, 2024.
https://doi.org/10.1007/978-981-97-4557-9_5

production conditions. The primary advantage of plant trials is their practical relevance; they provide a direct assessment of procedures and modifications under real operational parameters, making the results highly applicable to industrial practice. However, plant trials are not without drawbacks. They can be costly and disruptive to regular production, and the variability in operational conditions can lead to data consistency and replicability challenges. Additionally, the complexity and scale of steelmaking operations make it difficult to isolate specific variables, potentially impacting the clarity and applicability of the findings.

With the limitations associated with conducting plant trials, slag eye estimation in steelmaking ladles traditionally relied on physical and mathematical modeling approaches. Physical modeling involves scaled-down experimental setups replicating the ladle furnace conditions, allowing for a tangible investigation of slag behavior [3–5]. While this method offers a direct observation of the ladle hydrodynamics, its scalability to industrial conditions can be limited, often requiring significant simplifications and assumptions. On the other hand, mathematical modeling, particularly computational fluid dynamics (CFD), provides a more comprehensive understanding of the transport phenomena influencing the slag eye area [5–9]. Mathematical models offer the advantage of simulating various operational scenarios without the need for expensive physical experiments. However, they rely heavily on the accuracy of the input parameters and often demand significant computational resources. Both methods, while beneficial in understanding the fundamental aspects of slag behavior, face challenges in adaptability and precision when applied to the complex and variable conditions of real-world steelmaking processes.

Researchers have conducted studies to derive empirical correlations by meticulously analyzing physical and operational parameters in controlled settings, often employing scaled-down models or computational simulations [3–5]. Though grounded in empirical data, these correlations are not entirely accurate when applied to real-world steelmaking processes. They often require simplifications and assumptions that may not closely replicate complex physics during industrial operations. As a result, these empirical correlations do not accurately predict slag eye behavior in diverse scenarios. To address this issue, Conejo and Feng collected a robust dataset that consolidated all the published data from these prior investigations and developed empirical correlations to estimate slag eye area in single and dual-purged steelmaking ladles [10]. Although these correlations showed superior predictability than several existing correlations, they have fallen short in their accuracy, as indicated by the low coefficient of correlation values reported.

With the advent of digital transformation and smart manufacturing, there is a growing need for more accurate and dynamic methods to optimize steelmaking processes [11]. Applying machine learning (ML) techniques has gained significant attention in response to this need. ML can process complex datasets and extract meaningful patterns, thus providing a more accurate and adaptable approach to predicting slag eye area. This study leverages ML algorithms, such as Linear Regression, Support Vector Machine, and Artificial Neural Networks, to improve the predictability of the slag eye area in steelmaking ladles. Using these advanced computational techniques promises to enhance the predictability of slag eye area, leading to improved operational efficiency and process control in the steelmaking industry.

2 Methodology

This study employed the data from the same set of literature considered by Conejo and Feng from an extensive literature review spanning 25 years of research. The dataset comprises 13 input parameters that influence the slag eye area. These input parameters are tabulated in Table 1 below.

Table 1. Input parameters affecting the slag eye area.

S. No	Variable	Units
1	Liquid Bath Height	m
2	Slag Layer Thickness	m
3	Top diameter of the ladle	m
4	Bottom Diameter of the ladle	m
5	Diameter of the gas inlet nozzle	m
6	Radial position of gas inlet-1	m
7	Radial position of gas inlet-2	m
8	Distance between both the gas inlets	m
9	Density of the liquid bath	$\frac{kg}{m^3}$
10	Density of the top layer/slag	$\frac{kg}{m^3}$
11	Viscosity of the top layer/slag	$\frac{Kg}{ms}$
12	Gas inflow rate through inlet-1	*NLPM*
13	Gas inflow rate through inlet-2	*NLPM*

Figure 1 displays a collection of kernel density plots, which are used to visualize the distribution of continuous variables in a dataset. Kernel density plots are smooth, continuous versions of a histogram that can depict the density of data points over a continuous interval. From the image, we observe several distinct behaviors. Some plots display a monomodal distribution with a single peak, suggesting that most data points are concentrated around a central value. This is indicative of a single dominant factor within the variable being measured. For instance, variables like liquid bath height, top layer thickness, top diameter of the ladle, etc., exhibit this pattern, where most measurements are clustered around a median value, with frequency gradually decreasing as you move away from the peak. A few plots depict a multi-modal distribution, as seen in the bottom diameter of the ladle, the diameter of the gas inlet nozzle, etc. These plots have more than one peak, suggesting the presence of multiple distinct groups within the variable being measured, or a dual characteristic within the dataset.

All the plots in this figure depict right-skewed distributions. This means that for each of these variables, many observations are clustered towards the lower end of the scale, and there is a tail that extends to the right, indicating infrequent high-value observations. The right-skewed distributions in the data indicate a higher prevalence of physical modeling experimental results than large-scale industrial data. Moreover, the similarity in distribution shapes suggests that correlations or shared factors may influence these variables, which could be of interest for further statistical analysis or process optimization.

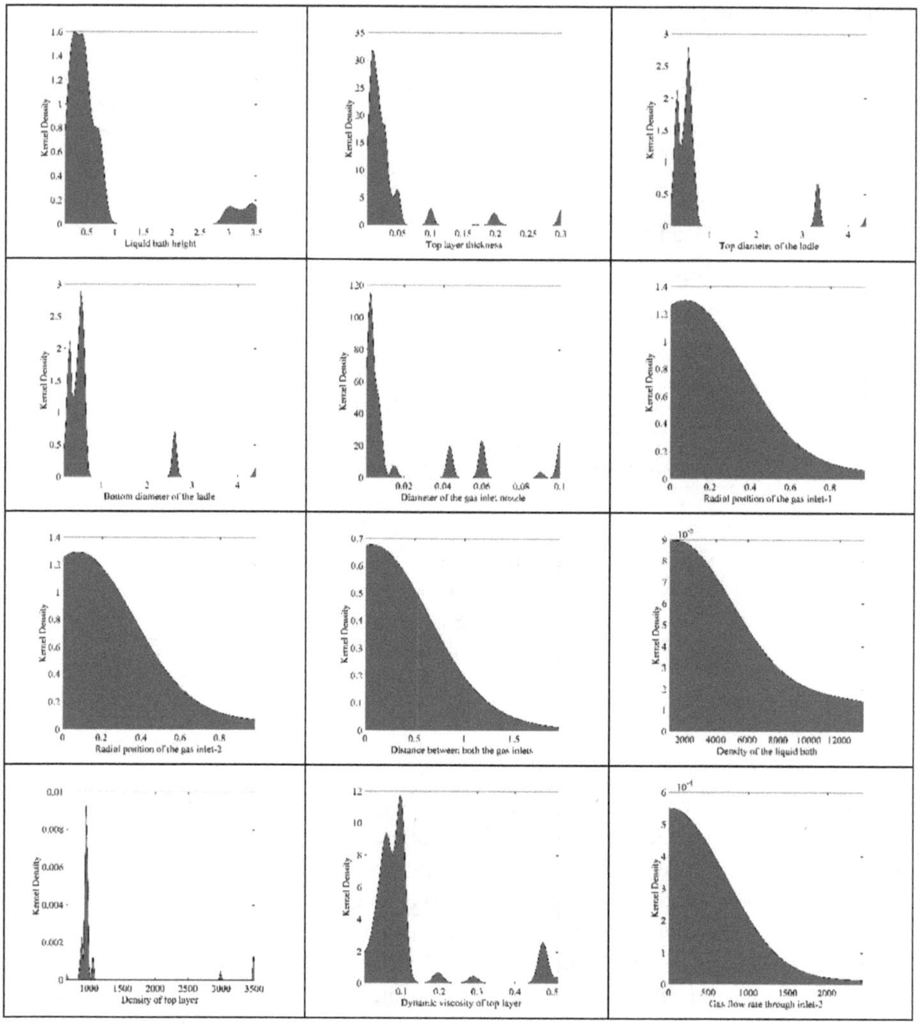

Fig. 1. Kernel density plots of the input variables.

Several machine learning models were employed to predict the slag eye area, focusing on comparing traditional linear regression, Support Vector Machines (SVMs), and Artificial Neural Networks (ANNs) with varying configurations. Linear regression serves as the baseline model for its simplicity and interpretability. It can also provide an empirical correlation relating the input parameters with the slag eye area. Consequently, an SVM with a linear kernel function was also considered, aiming to capture non-linear relationships without the complexity of non-linear kernels. Unlike linear regression, SVM seeks to find the best margin (distance between the line and the nearest points of all the classes, also known as support vectors) that separates different classes in the feature space. Furthermore, ANNs consist of layers of interconnected nodes or neurons, where each connection has a weight associated with it. These networks can model complex non-linear relationships between inputs and outputs through a process known as training, where the weights are adjusted to minimize the difference between the actual and predicted outputs. In the current study, ANNs with configurations ranging from a single layer to a three-layered architecture were also considered for testing the slag eye area predictability. To ensure the models' robustness and generalizability, a 5-fold cross-validation approach was adopted. Additionally, 10% of the data was set aside as a test set to provide an unbiased evaluation of the model performance on unseen data.

In summary, while linear regression is straightforward and used for predicting a continuous outcome, SVM with a linear kernel provides a more robust alternative for classification by maximizing the margin between classes. In contrast, ANNs offer a highly flexible approach capable of capturing complex non-linear relationships through a network of neurons, making them suitable for a wide range of applications beyond linear contexts. Each model's effectiveness and suitability depend on the nature of the data and the specific task at hand.

3 Results and Discussion

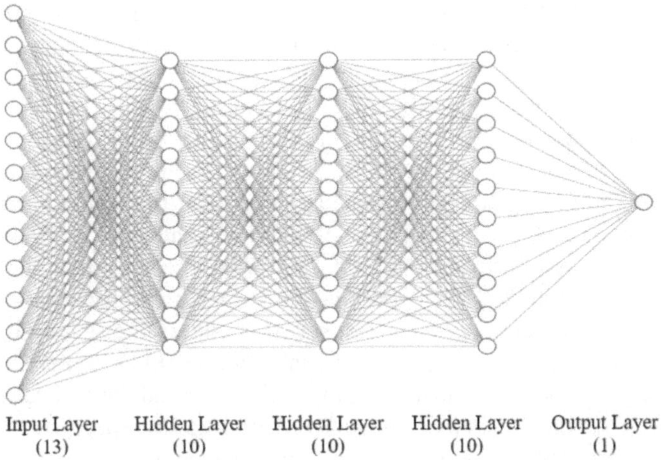

Fig. 2. Network map of the tri-layered neural network.

Figure 2 shows the network map of a tri-layer neural network used in the current study. The image depicts a feedforward neural network architecture consisting of an input layer, three hidden layers, and an output layer. The input layer has 13 nodes, corresponding to 13 different input features that the network uses to make its predictions. Each of the three hidden layers has 10 nodes, which are units that perform computations and transfer information through the network. The connections between nodes, represented by lines, indicate the flow of data and the network's learned weights that are adjusted during training. The final output layer has a single node, suggesting that the network is designed to perform a regression task with a single continuous output value, i.e., the slag eye area.

Table 2. Summary of machine learning model performance.

Model Type	R^2 (Validation)	R^2 (Testing)	Remarks
Linear Regression	0.83	–	All data used for linear least square analysis
SVM	0.82	0.79	Linear kernel function
ANN	0.93	0.94	1 layer, 10 nodes
ANN	0.94	0.90	1 layer, 100 nodes
ANN	0.95	0.94	2 layers, 10 nodes each
ANN	0.96	0.96	3 layers, 10 nodes each

The evaluation of the machine learning models yielded insightful results, summarized in Table 2. The linear regression analysis could not predict the slag opening area satisfactorily. In addition, the regression analysis performed by Conejo and Feng on the same dataset yielded an R^2 value of 0.76 for single-purged ladles and 0.90 for dual-purged ladles, which are also poor. The three-layer ANN with 10 nodes in each layer emerged as the most accurate model, achieving a goodness of fit estimate value (R^2) of 0.96 for validation and an impressive 0.96 for the test set. This indicates a high level of predictive accuracy, suggesting that the model can generalize well to new data. The superior performance of the three-layer neural network can be attributed to its deeper architecture, which is adept at capturing the complex, non-linear relationships between the input parameters and the slag eye area. The depth of the network allows for a more nuanced understanding of the process, leading to more precise predictions. Compared with the linear regression model, the three-layer neural network substantially improved predictive accuracy, particularly for the test set. This underscores the limitations of linear models when dealing with complex industrial processes like steelmaking. Moreover, while the SVM performed reasonably well in validation, it did not maintain its accuracy in the test set. This could indicate the model's sensitivity to the specific data distribution of the validation fold and its consequent inability to generalize.

Figure 3 illustrates the training and validation loss of the tri-layered neural network over numerous epochs, which are full iterations over the entire dataset. The training loss (in blue) and the validation loss (in orange) both demonstrate a sharp decrease initially, which indicates that the model is quickly learning from the data. As the epochs increase, both losses level off, converging to a stable, low error rate, suggesting that the model is generalizing well and not overfitting the training data. A significant point to note is that the validation loss closely tracks the training loss, a sign of the model's ability to

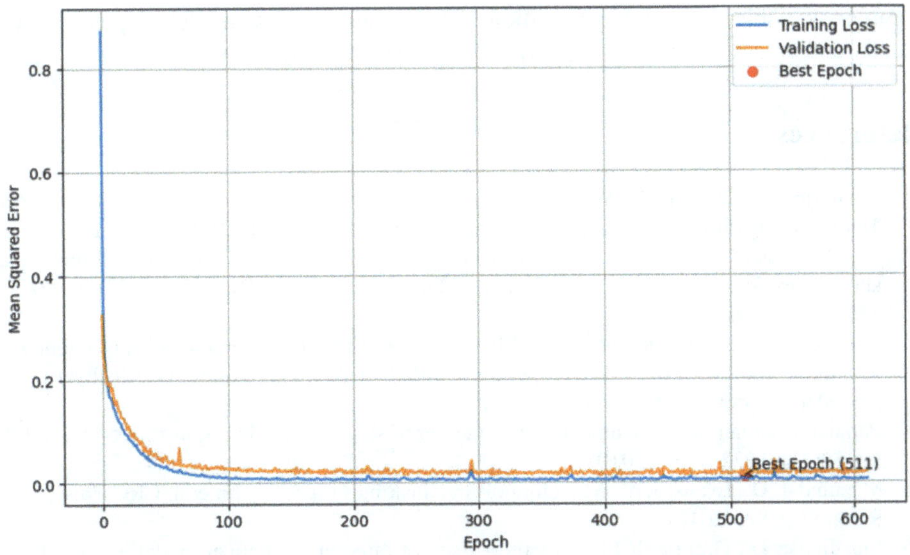

Fig. 3. Training and validation loss for the tri-layered neural network.

perform consistently on unseen data. The "Best Epoch" marker at epoch 511 indicates the epoch at which the validation loss was minimal, signifying the optimal stopping point for training before the model overfits or before the improvements become negligible. The graph shows a successful training process with the model achieving a low mean squared error, which is desirable for prediction accuracy in regression tasks.

The findings of this study have significant practical implications. Applying a three-layer neural network for predicting the slag eye area could significantly enhance operational efficiency in steelmaking. It offers a real-time monitoring and control tool, potentially leading to optimized inclusion removal and reduced process variability. While the results are promising, it is important to consider the limitations of this study. The predictive models rely on the quality and range of the data collected. Since the dataset originates from diverse sources, underlying biases or inconsistencies may not be accounted for in the model training process. Additionally, the models, particularly the neural networks, require a substantial volume of data for training, which may not be readily available in some industrial contexts.

4 Conclusions

This study applies machine learning models to predict the slag eye area, using a comprehensive 13-feature dataset from literature spanning 25 years. Various machine learning models were tested, including Linear Regression, Support Vector Machines, and Artificial Neural Networks. The three-layer Artificial Neural Network with 10 nodes in each layer proved to be the most accurate model for predicting slag eye area. Unlike empirical correlations, the machine learning approach offers a dynamic method that can adapt to complex, variable steelmaking conditions. The success of these models depends on the quality and breadth of data, especially those on industrial systems.

References

1. Chatterjee, S., Chattopadhyay, K.: Physical modeling of slag 'eye' in an inert gas-shrouded Tundish using dimensional analysis. Metall. Mater. Trans. B **47**(1), 508–521 (2016)
2. Liu, Y., Ersson, M., Liu, H., Jonsson, P.G., Gan, Y.: A review of physical and numerical approaches for the study of gas stirring in ladle metallurgy. Metall. Mater. Trans. B **50**, 555–577 (2019)
3. Castillejos, A.H., Brimacombe, J.K.: Measurement of physical characteristics of bubbles in gas-liquid plumes: Part II. Local properties of turbulent air-water plumes in vertically injected jets. Metall. Trans. B **18**, 659–671 (1987)
4. Peranandhanthan, M., Mazumdar, D.: Modeling of slag eye area in argon stirred ladles. ISIJ Int. **50**(11), 1622–1631 (2010)
5. Mazumdar, D., Evans, J.W.: Modeling of Steelmaking Processes, 1st edn. CRC Press, Boca Raton FL USA (2010)
6. Mazumdar, D., Guthrie, R.I.L.: A Comparison of three mathematical modeling procedures for simulating fluid flow phenomena in bubble-stirred ladles. Metall. Mater. Trans. B **25**, 308–312 (1994)
7. Mantripragada, V.T., Sarkar, S.: Study of transient behavior of slag layer in bottom purged ladle: a CFD approach. In: Mason, P., et al. (eds.) Proceedings of the 4th World Congress on Integrated Computational Materials Engineering (ICME 2017). TMMMS, pp. 145–153. Springer, Cham (2017). https://doi.org/10.1007/978-3-319-57864-4_14
8. Mantripragada, V.T., Sarkar, S.: Slag eye formation in single and dual bottom purged industrial steelmaking ladles. Can. Met. Quarterly **59**(2), 159–168 (2020)
9. Mantripragada, V.T., Sarkar, S.: Multi-objective optimization of bottom purged steelmaking ladles. Trans. Indian Inst. Met. **75**, 2289–2298 (2022)
10. Conejo, A.N., Feng, W.: Ladle eye formation due to bottom gas injection: a reassessment of experimental data. Metall. Mater. Trans. B **53**, 999–1017 (2022)
11. Palovaara, T., Visuri, V.V., Fabritius, T.: Physical modelling of gas injection in a ladle. In: 7th International Congress on Science and Technology of Steelmaking: The Challenge of Industry 4.0 2018, pp. 1154–1163. Associazione Italiana di Metallurgia (2018)

An Overview of Sustainable Solutions Towards Recycling of Metallurgical Slags

Navneet Singh Randhawa(✉)

Metal Extraction and Recycling Division, CSIR-National Metallurgical Laboratory, Jamshedpur 831003, India
nsr@nmlindia.org

Abstract. Metallurgical slags, such as copper slag, electric arc furnace (EAF) slag, and basic oxygen furnace (BOF) slag, are byproducts of metal production processes. Traditionally seen as waste, these slags possess significant potential as a resource for metals and valuable products. Recycling technologies and energy-efficient methods can extract critical metals from slags. Innovative techniques enable the extraction of metals from copper slag and the use of secondary waste in construction materials. Energy-efficient processes convert EAF slag into valuable products and recover iron from BOF slag for steel production. Efficient slag recycling promotes energy savings and sustainable resource management, contributing to a sustainable future. By employing innovative technologies and energy-efficient methods, the potential of slags can be harnessed while minimising waste generation and environmental impact. Embracing these recycling opportunities contributes to a sustainable future and efficient resource management.

Keywords: Copper slag · EAF slag · BOF slag · Resource · Recycling

1 Introduction

Metallurgical slag is a by-product of the metal extraction process and plays a significant role in various industrial applications. It is a secondary raw material to improve metal extraction practices [1]. Metallurgical slags are produced from the combination of various materials, such as lime, silica, and alumina, and their properties are crucial in high-temperature metallurgical processes [2]. The dissolution rate of different types of metallurgical lime into molten slag has been studied, highlighting the importance of understanding the behaviour of individual components in slag formation [3]. Furthermore, the relationship between the viscous properties of oxide melts and the slag structure has been extensively investigated, emphasising the importance of understanding the structural properties of slags in ironmaking [4]. Metallurgical slags have also found applications in the construction industry, where they are used as alternative materials for producing sustainable concrete. Additionally, they are utilised to produce controlled low-strength materials, with blast-furnace (BF) slag being a commonly used supplementary cementitious material due to its high hydration activity [5, 6]. Moreover, metallurgical slags, such as steel slag, copper slag, lead-zinc slag, and electric furnace ferronickel slag,

A. Tiwari et al. (Eds.): iConFIRM 2023, SPPHY 308, pp. 47–56, 2024.
https://doi.org/10.1007/978-981-97-4557-9_6

are commonly used as aggregates and supplementary cementitious materials and, further underlining their versatility in various industrial applications [7]. The environmental impact of metallurgical slags has been a subject of study, with concerns raised about their accumulation in heaps over the years, posing challenges for the metallurgical industries. However, from an industrial symbiosis perspective, the addition of metallurgical slags has shown promise in obtaining green friction materials with reduced emissions, highlighting their potential to contribute to sustainable practices [8, 9]. In addition to their industrial applications, metallurgical slags have been studied for their potential adverse effects when used as supplementary cementitious materials and aggregates, emphasising the importance of understanding their impact on the properties of construction materials [7]. Furthermore, the hazardous nature of old metallurgical slags due to their potentially toxic elements has been highlighted, emphasising the need for proper disposal and management of these materials [10]. Nevertheless, metallurgical slag is a versatile by-product of metal extraction processes with diverse applications in various industries, including construction, steelmaking, and environmental sustainability. Understanding its properties, behaviour, and potential impact is crucial for maximising its benefits while mitigating potential adverse effects [11]. This review entails the recent advances in the area of metallurgical slag utilization in brief. A discussion on the economic aspects of the recycling of slag is given, emphasizing the end use. Efforts on the lab scale and on the industrial scale are highlighted for a brief overview of the recent endeavors in this direction.

2 Recent Advances in Slag Recycling

Recent advances in slag recycling have indeed contributed significantly to sustainable industrial practices. For instance, innovative recycling routes have been developed to meet the increasing demand for battery raw materials, ensuring future sufficiency. Additionally, the utilisation ratios of ironmaking and steelmaking slags have risen steadily, driven by technological advances in Japan and China. Unprecedented technological advances have led to the thermal isolation of clean alloys from waste slag and polymeric residue of electronic waste, addressing the challenges posed by the considerable amount of electronic waste generated. Furthermore, recent years have seen the proposal of a new refining method to increase the utilisation of CaO or dephosphorisation ability of slags by using a solid/liquid coexisting multiphase slag in Japan. The recycling of copper smelter slag for the production of high-value-added products and its use in cement-based electromagnetic interference composites has been explored, addressing concerns about the overcapacity of the cement industry and the urgent need for alternative slag utilisation. Moreover, the marketable products derived from blast furnaces and steelmaking slags have been used in various applications, including cement, roadbed material, ground improvement material, civil engineering material, and fertiliser, with only a tiny percentage processed as waste in landfills. In the construction industry, recent studies have focused on the compressive strength, chloride ion penetrability, and carbonation characteristics of concrete with mixed slag aggregate, addressing the shortage of natural aggregates and promoting sustainable development. Additionally, the physicochemical characteristics of silico manganese slag have been systematically reviewed, with successful processing into various construction materials and manganese extracts. Advancements in computational modeling and extensive data mining have been applied to predict the viscosity of slags, impacting metallurgical efficiency and sustainable development in high-temperature industries. Applying steel slag powder and recycled aggregates in self-compacting concrete has been introduced to improve sustainability performance. Furthermore, the dynamic shear properties of recycled waste steel slag used as a geo-backfill material have been studied, reflecting the growing trend of utilising industrial solid wastes in civil engineering [12]. Table 1 summarizes some of the recent studies on utilization of slags. In environmental sustainability, recent research has focused on the environmental assessment of recycling steelmaking slag, aiming to ensure safe recycling and address challenges related to weaknesses in the recycling system [13]. Additionally, the reuse of silico-manganese slag has been explored to meet the growing demand for raw manganese and develop economic recovery methods from secondary sources [14, 15].

As a result of continuous efforts, slag recycling has encompassed a wide range of applications, from addressing electronic waste challenges to improving the sustainability performance of construction materials. For instance, using metallurgical waste slag as a cementitious material in controlled low-strength materials (CLSMs) is crucial to sustainable construction industry development [16]. Additionally, the viscosity of slag significantly impacts metallurgical efficiency and recycling, thereby affecting the quality of metal and the sustainable development of high-temperature industries [17]. Environmental and socioeconomic impacts of metallurgical slags, such as copper slag, emphasise the significance of recycling to mitigate the accumulation of slag heaps, thereby reducing environmental burdens and promoting sustainable waste management practices [18]. Moreover, integrating steel slags in asphalt concretes as recycled aggregates reduce greenhouse gas emissions and resource consumption, aligning with the principles of industrial symbiosis and sustainable material utilisation [19]. Using graphite-rich fractions from spent lithium-ion batteries as an alternative reductant in nickel slag cleaning demonstrates the potential for integrating recycling processes with existing metallurgical practices, contributing to resource efficiency and sustainable metal production [14]. Additionally, the reuse of silico-manganese slag aligns with the EU's ecological economy support program, emphasising the importance of recovering and recycling resources from metallurgical wastes for sustainable development [20]. In environmental sustainability, recent research has focused on the environmental assessment of recycling steelmaking slag, aiming to ensure safe recycling and address challenges related to weaknesses in the recycling system [21]. These advancements reflect the ongoing efforts to optimise waste management processes, enhance resource efficiency, and promote sustainable industrial practices.

3 Economic Implications

The economic implications of slag recycling are multifaceted and have been the subject of extensive research. The recycling and reutilization of materials have been shown to increase the efficiency of industrial processes, thereby enhancing the economic profitability of the industrial sector [22]. Furthermore, the economic and environmental

Table 1. Some recent studies on the use of metallurgical slags

Type of Slag	Origin	Usage	Reference
Steel slag	BOF	Iron recovered by carbothermic reduction	[14]
Silicomanganese slag	Submerged arc furnace producing ferroalloys	Construction material, e.g. Bricks, cement, geopolymers	[15]
Blast furnace and ferronickel slag	Ironmaking and ferroalloy industry	Used as mixed aggregates for concrete	[16]
Assorted slag components	All the metallurgical slag types	Viscosity modeling for predicting and supporting the end use by recycling	[17]
Blast furnace slag and EAF slag	Granulated blast furnace slag and dumped BOF slag	A mixture of slag is used as a soil conditioner to mitigate acidic soil	[18]
EAF slag	Alloy steel making	EAF slag used in asphalt replacing natural stones	[19]
Silicomanganese slag	Submerged arc smelting	Manganese recovered, and the resulting material used in construction material	[20]
Steelmaking slag	BOF process	Toxicity criteria of the slag studies towards safe disposal and utilization in construction materials	[21]
Copper slag	Flash smelter	Copper and magnetite recovered	[22]
Steel slag	BOF process	Use as aggregate in asphalt mixture	[23]

benefits can be realised by the circular utilisation of reclaimed asphalt pavement (RAP), which has received extensive attention for contributing to sustainable practices in the construction industry [23]. Using steel slag as aggregate in pile composite foundations has been demonstrated to bring significant economic and environmental benefits [24]. Additionally, incorporating spent sand into clay bricks has been shown to reduce the costs of brick manufacturing and waste sand treatment, thereby offering economic advantages [25]. The bulk use of steel slag as fine aggregate can save the overconsumption of natural resources and decrease environmental pollution, leading to economic benefits [26]. Moreover, recycling residual integrated gasification combined cycle (IGCC) slags has been shown to have both environmental and economic benefits, improving the technological properties of ceramics [27]. The reuse of silico-manganese slag has been aligned with the sustainable use of natural resources for sustainable development, highlighting the economic benefits of recycling metallurgical slags. Additionally, recycling valuable metals from waste electrical and electronic equipment (WEEE) slag is vital in establishing a circular economy, contributing to economic sustainability [28]. Furthermore, using metallurgical slags as aggregate in concrete has been aimed at enhancing the "4R-reduce, reuse, recycle, and renew strategy," optimising waste management processes [19]. These findings collectively underscore the economic benefits of slag recycling across various industrial sectors, emphasising its potential to contribute to sustainable and cost-effective practices.

4 Some Lab Scale Studies

Tan et al. [29] study focuses on the efficient centrifugal granulation of molten blast furnace slag but does not provide evidence of industrial-scale applications. - Kholkina et al. [30] investigated the influence of alkaline pretreatment, synthesis time, and temperature on synthesising and characterising novel catalytic materials using industrial slag on a lab scale. The capacity of tundish slag to absorb inclusions was investigated by Pereira et al. [31]. They used the calcium aluminate-based covering material combined with rice hull ash to investigate the slag's absorbing capacity towards inclusion from the metal phase. Taskinen et al. (2013) studied the freeze-lining formation and microstructure in the flash smelter slag. The study indicated 850 °C as suitable temperature to obtain freeze-lining matrix [32]. A study by Liu et al. (2019) provided insights into the granulation of a modified basic oxygen furnace (BOF) steel slag [33]. In this study, an Al_2O_3- and SiO_2-modified BOF slag was water-granulated and investigated for its crystallisation behaviour during the granulation process. The critical cooling rate was determined through the Time-Temperature-Transformation (TTT) diagrams constructed using in-situ confocal scanning laser microscopy. Lin et al. (2020) proposed multi-phase slag refining as an effective way to improve the utilisation of solid lime flux and the efficiency of dephosphorisation in steelmaking practices [34]. This proposed concept employs the coexisting state of both liquid and solid phases in refining. The solid-liquid coexisting state of the slag changes significantly by lime dissolution and increasing temperature, as well as FeO content in the molten slag. Numerous lab-scale studies investigated the application of slags in the construction material. The optimisation of blended mortars using steel slag sand was carried out to prove the potential use of slag to replace natural sand [35]. The study used cement clinker, steel slag sand, GGBFS and gypsum. It was indicated that applying steel slag sand may reduce the dosage of clinker and increase the content of industrial waste products such as GGBFS with improved mechanical performance and microstructural characteristics. Another study on using slag as an aggregate was attempted by Nadeem and Profale (2012) [36], applying Taguchi's design of experiments. According to the study, concrete compressive strength increased by 5% and 7% for nearly all % of regular crushed coarse aggregate replacements with crystallised aggregates. The study also found that complete substitution by slag aggregate for normal crushed coarse improved flexure strength by 6–8% for all replacements and 30–50% for fine and coarse aggregates with slag. Ling et al. conducted an industrial-scale study on the stabilisation of BOF slag utilising hot-stage carbonation treatment [37].

5 Industrial Scale Endeavours

This section provides a few industrial-scale innovative attempts to recycle various slag. The industrial-scale treatment processes of steel slag have been aimed at stabilisation to improve its use in road making, cement manufacturing, and other building materials. An innovative waste heat recovery study was conducted in China [38] (Table 2).

The airflow and the thickness of steel slag influenced the cooling air temperature change and the waste heat recovery. The adequate heat recovered by air from 1 kg of steel slag reached approximately 1000 MJ. The heat recovery rate of the slag was up to 74.4% when the heat slag in the slag tank was 400 mm (about 3.96 tons), and the blast volume was 14687 Nm^3/h. 28-day activity index of steel slag is similar to the original one. The most important aspect of this process is that it can be integrated fully into the existing high-tech steel slag treatment in China. A study by Davydenko et al. (2014) used briquettes to obtain a combined slag foaming and waste product reduction in the EAF process [39]. Today, two of the most critical roles in sustainable steelmaking are the efficient use of the energy required to produce stainless steel (SS) in the electric arc furnace, slag foaming, and recycling of waste products. The study was conducted in Sweden on an industrial scale in 85-ton EAF to evaluate the foaming tendency of slag in the EAF process during stainless steel production. The slag's basicity and density of the briquette significantly affected the foaming process. The highest Slag Foaming Rate (SFR) value in this study was obtained for the EAF slags with basicity in the range of 1.31–1.49, having a melting temperature of about 1500–1600ºC. Coal ash-modified BF slag was initially investigated on a laboratory scale by a high-speed air injection method to obtain slag wool with appropriate mechanical properties. Subsequently, a 20,000 tons per annum slag wool plant was constructed in Shanxi Province, China [40]. The raw materials were BF slag and coal ash. A reduction 70% in energy consumption and a decrease of 90% in pollution emissions compared to traditional methods of mineral-wool manufacturing were recorded. The use of Waelz Slag (WS), generated in EAF dust recycling, in clay bricks was studied in a pilot scale extruder (Verdés, 050-C) to obtain rectangular shape (150 mm × 30 mm × 20 mm) samples [40]. The green samples were dried in an industrial oven and fired in an industrial-scale kiln up to 1020 °C. It was found that the blending of WS in clay bricks can be scaled up for additions of WS $\leq$ 10%. Recently, a six-lane public road with a kilometre-long stretch was built in Hazira Industries, Gujrat state of India. The joint venture of Central Road Research Institute (CRRI), India, government think-tank NITI Ayog, the Union Ministry of Steel, and ArcelorMtttal-Nippon Steel (AM/NS) endeavoured this development [42]. It is estimated that the general construction cost while using a processed steel slag per square meter of road is Rs 1,150, which is lower than for a bitumen road (Rs 1,300) and for a cement or a concrete one (Rs 2,700). The carbon footprint for steel slag roads is much lower than those built using natural aggregates, which require processing before road making. Steel slag road does not need blasting, drilling, or crushing as required for natural materials.

Overall, the utilization of slags spans a wide range of applications, from enhancing concrete properties to serving critical roles in metallurgical processes and environmental remediation efforts. Understanding the composition and behavior of different slags is crucial for optimizing their use in various industries.

Table 2. Some extensive/industrial-scale innovative attempts to recycle various slag.

Sl no	Type of slag	Scale	Country	Remarks	Reference
1	BOF slag	3.96 tons	China	Adequate heat recovered up to ~1000 MJ/kg steel slag by air The heat recovery rate was as high as 74.4%	[38]
2	SS slag	85 ton EAF at Outokumpu Stainless	Avesta, Sweden	10 heats were carried out to observe the foaming behaviour of slag using light and heavy briquettes	[39]
3	BF slag	20,000 tons/annum	Shanxi Province, China	A 90% pollution emission decrease and 70% reduction in energy consumption compared to traditional methods of Mineral-wool manufacturing	[40]
4	Waelz slag (WS)	Tested in industrial scale drier and brick kiln in Spain	Santander, Spain	WS in clay bricks was used up for additions of WS $\leq$ 10 wt% having desired mechanical properties compared to industrial bricks	[41]
5	BOF slag	Tested for 1 km road to replace natural aggregates	Gujrat, India	Using steel slag in bitumen road building has significant environmental and economic benefits	[42]

References

1. Brandaleze, E., Benavidez, E., Santini, L.: Treatments and recycling of metallurgical slags. In: Recovery and Utilization of Metallurgical Solid Waste, pp. 35–52. IntechOpen, London (2018)
2. Ri, S., Ryom, K., Ra, J.: Derivation of CaO-SiO2-Al2O3 system slag viscosity equation by GP. J. Min. Metall. Sect. B. **57**(3), 309–317 (2021). https://doi.org/10.2298/jmmb190919027r
3. Maruoka, N., Ishikawa, A., Shibata, H., Kitamura, S.Y.: Dissolution rate of various limes into steelmaking slag. High Temp. Mater. Processes **32**(1), 15–24 (2013). https://doi.org/10.1515/htmp-2012-0049
4. Sohn, I., Min, D.J.: A review of the relationship between viscosity and the structure of calcium-silicate-based slags in ironmaking. Steel Res. Int. **83**(7), 611–630 (2012). https://doi.org/10.1002/srin.201200040
5. Collivignarelli, M.C., et al.: The production of sustainable concrete with the use of alternative aggregates: a review. Sustainability **12**(19), 7903 (2020). https://doi.org/10.3390/su12197903
6. Liu, Y., Su, Y., Xu, G., Chen, Y., You, G.: Research progress on controlled low-strength materials: metallurgical waste slag as cementitious materials. Materials **15**(3), 727 (2022). https://doi.org/10.3390/ma15030727
7. Zhao, Q., Pang, L., Wang, D.: Adverse effects of using metallurgical slags as supplementary cementitious materials and aggregate: a review. Materials **15**(11), 3803 (2022). https://doi.org/10.3390/ma15113803
8. Gabasiane, T., Danha, G., Mamvura, T., Mashifana, T., Dzinomwa, G.: Environmental and socioeconomic impact of copper slag—a review. Crystals **11**(12), 1504 (2021). https://doi.org/10.3390/cryst11121504
9. Jayashree, P., Matějka, V., Leonardi, M., Straffelini, G.: A novel path towards limiting non-exhaust particulate matter emissions of a commercial friction material through the addition of metallurgical slag. Sci. Rep. **13**(1), 666 (2013). https://doi.org/10.1038/s41598-023-27932-6
10. Zolotova, E.: Studies of soils and vegetation on non-ferrous metallurgy slag dumps. Int. J. Bio-resour. Stress Manag. **12**(1), 40–46 (2021). https://doi.org/10.23910/1.2021.2178a
11. Chandel, S.S., Singh, P.K., Katiyar, P.K., Randhawa, N.S.: A review on environmental concerns and technological innovations for the valorization of steel industry slag. Mining Metall. Explor. **40**(6), 2059–2086 (2023). https://doi.org/10.1007/s42461-023-00886-z
12. Ruismäki, R., Dańczak, A., Klemettinen, L., Taskinen, P., Lindberg, D., Jokilaakso, A.: Integrated battery scrap recycling and nickel slag cleaning with methane reduction. Minerals **10**(5), 435 (2020)
13. Matsuura, H., Yang, X., Li, G., Yuan, Z., Tsukihashi, F.: Recycling of ironmaking and steelmaking slags in Japan and China. Int. J. Miner. Metall. Mater. **29**(4), 739–749 (2022). https://doi.org/10.1007/s12613-021-2400-5
14. Khayyam Nekouei, R., Maroufi, S., Assefi, M., Pahlevani, F., Sahajwalla, V.: Thermal isolation of a clean alloy from waste slag and polymeric residue of electronic waste. Processes **8**(1), 53 (2020). https://doi.org/10.3390/pr8010053
15. Nath, S.K., Randhawa, N.S., Kumar, S.: A review on characteristics of silico-manganese slag and its utilization into construction materials. Resour. Conserv. Recycl. **176**, 105946 (2022)
16. Choi, S.J., Kim, Y.U., Oh, T.G., Cho, B.S.: Compressive strength, chloride ion penetrability, and carbonation characteristic of concrete with mixed slag aggregate. Materials **13**(4), 940 (2020). https://doi.org/10.3390/ma13040940
17. Huang, A., Huo, Y., Yang, J., Gu, H., Li, G.: Computational modeling and prediction on viscosity of slags by big data mining. Minerals **10**(3), 257 (2020). https://doi.org/10.3390/min10030257

18. Ghisman, V., Muresan, A.C., Buruiana, D.L., Axente, E.R.: Waste slag benefits for correction of soil acidity. Sci. Rep. **12**(1), 16042 (2022). https://doi.org/10.1038/s41598-022-20528-6
19. Vaiana, R., Balzano, F., Iuele, T., Gallelli, V.: Microtexture performance of EAF slags used as aggregate in asphalt mixes: a comparative study with surface properties of natural stones. Appl. Sci. **9**(15), 3197 (2019) https://doi.org/10.3390/app9153197
20. Buruiana, D.L., Obreja, C.D., Herbei, E.E., Ghisman, V.: Re-use of silico-manganese slag. Sustainability **13**(21), 11771 (2021). https://doi.org/10.3390/su132111771
21. Lee, M.Y., Kang, J.H., Hwang, D.G., Yoon, Y.S., Yoo, M.S., Jeon, T.W.: Environmental assessment of recycling (EAoR) for safe recycling of steelmaking slag in the Republic of Korea: applications, leaching test, and toxicity. Sustainability **13**(16), 8805 (2021). https://doi.org/10.3390/su13168805
22. Fernández-González, D., Prazuch, J., Ruiz-Bustinza, Í., González-Gasca, C., Gómez-Rodríguez, C., Verdeja, L.F.: Recovery of copper and magnetite from copper slag using concentrated solar power (CSP). Metals **11**(7), 1032 (2021). https://doi.org/10.3390/met11071032
23. Wang, Z., et al.: Quantitative assessment of road performance of recycled asphalt mixtures incorporated with steel slag. Materials **15**(14), 5005 (2022). https://doi.org/10.3390/ma15145005
24. Zhang, B., Cui, S.: Field tests of cement fly-ash steel-slag pile composite foundation. J. Test. Eval. **45**(3), 860 (2016). https://doi.org/10.1520/jte20140422
25. Alonso-Santurde, R., et al.: Technological behaviour and recycling potential of spent foundry sands in clay bricks. J. Environ. Manage. **92**(3), 994–1002 (2011). https://doi.org/10.1016/j.jenvman.2010.11.004
26. Wu, K., et al.: Quantitative evaluation of interfacial transition zone of sustainable concrete with recycled and steel slag as aggregate. Struct. Concr. **22**(2), 926–938 (2021). https://doi.org/10.1002/suco.202000135
27. Martín, I.I., Echeverría, A.A., García-Romero, E.: Recycling of residual IGCC slags and their benefits as degreasers in ceramics. J. Environ. Manag. **129**, 1–8 (2013). https://doi.org/10.1016/j.jenvman.2013.05.055
28. Buchmann, M., et al.: Evaluation of recyclability of a WEEE slag by means of Integrative X-ray computer tomography and SEM-based image analysis. Minerals **10**(4), 309 (2020). https://doi.org/10.3390/min10040309
29. Tan, Y., Zhu, X., Wang, H., Lv, Y.W., Ding, Y.D., Liao, Q.: A novel stacked rotary cup atomizer toward efficient centrifugal granulation of molten blast furnace slag. Steel Res. Int. **92**(11), 2100207 (2021). https://doi.org/10.1002/srin.202100207
30. Kholkina, E., et al.: Synthesis and characterization of novel catalytic materials using industrial slag: influence of alkaline pretreatment, synthesis time and temperature. Top. Catal. **62**, 738–751 (2019). https://doi.org/10.1007/s11244-019-01162-5
31. Pereira, M.M.D.S.M., et al.: Tundish slag capacity to absorb inclusions when using calcium aluminate based covering material combined with rice hull ash (2016). https://doi.org/10.5151/1982-9345-27959
32. Taskinen, P., Kaskiala, M., Miettinen, K., Jansson, J.: Freeze-lining formation and microstructure in a direct-to-blister flash smelting slag. J. Manuf. Sci. Prod. **13**(1–2), 77–83 (2013). https://doi.org/10.1515/jmsp-2012-0024
33. Liu, C., Lopez Gonzalez, P.L., Huang, S., Pontikes, Y., Blanpain, B., Guo, M.: Experimental and mathematical simulation study on the granulation of a modified basic oxygen furnace steel slag. Metallurg. Mater. Trans. B **50**, 1260–1268 (2019). https://doi.org/10.1007/s11663-019-01543-x
34. Lin, W., Jiao, S., Zhou, K., Sun, J., Feng, X., Liu, Q.: A review of multi-phase slag refining for dephosphorization in the steelmaking process. Front. Mater. **7**, 602522 (2020). https://doi.org/10.3389/fmats.2020.602522

35. Chen, M., Zhou, M., Wu, S.: Optimization of blended mortars using steel slag sand. J. Wuhan Univ. Technol.-Mater. Sci. Ed. **22**(4), 741–744 (2007). https://doi.org/10.1007/s11595-006-4741-3
36. Mohammed, N., Arun D.P.: Utilization of industrial waste slag as aggregate in concrete applications by adopting Taguchi's approach for optimization. Open J. Civil Eng. 02(03), 96–105 (2012). https://doi.org/10.4236/ojce.2012.23015
37. Santos, R.M., et al.: Stabilization of basic oxygen furnace slag by hot-stage carbonation treatment. Chem. Eng. J. **203**, 239–250 (2012). https://doi.org/10.1016/j.cej.2012.06.155
38. Zhang, T., Qiu, G., Wang, H., Guo, M., Cheng, F., Zhang, M.: In-suit industrial tests of the highly efficient recovery of waste heat and reutilization of the hot steel slag. ACS Sustain. Chem. Eng. **9**(10), 3955–3962 (2021). https://doi.org/10.1021/acssuschemeng.1c00081
39. Davydenko, A., Karasev, A., Lindstrand, G., Jönsson, P.: Investigation of slag foaming by additions of briquettes in the EAF during stainless steel production. Steel Res. Int. **86**(2), 146–153 (2015). https://doi.org/10.1002/srin.201400036
40. Zhao, D., Zhang, Z., Tang, X., Liu, L., Wang, X.: Preparation of slag wool by integrated waste-heat recovery and resource recycling of molten blast furnace slags: from fundamental to industrial application. Energies **7**(5), 3121–3135 (2014). https://doi.org/10.3390/en7053121
41. Cifrian, E., Coronado, M., Quijorna, N., Alonso-Santurde, R., Andrés, A.: Waelz slag-based construction ceramics: effect of the trial scale on technological and environmental properties. J. Mater. Cycles Waste Manag. **21**, 1437–1448 (2019). https://doi.org/10.1007/s10163-019-00896-4
42. E-Paper. The Indian Express, Surat gets India's first steel slag road: What is it, how is it different from regular ones? https://indianexpress.com/article/explained/surat-indias-first-steel-slag-road-7847675

Structural Materials and Mechanical Metallurgy

An Investigation on Mechanical and Electrochemical Properties of Ti-6Al-4V alloy by Scheduling Heat-treatment in Pure β and α + β region

Bishnu Prasad Mahto[1(✉)], Daljeet Singh[2], Rajiv Kumar[2(✉)], Rajesh Kumar Rai[1(✉)], and M. K. Mishra[1(✉)]

[1] Department of Metallurgical and Materials Engineering, Malaviya National Institute of Technology Jaipur, Rajasthan 302017, India
{2021rmt9086,rajesh.meta,manjesh.meta}@mnit.ac.in

[2] Department of Metallurgical and Materials Engineering, Indian Institute of Technology Ropar, Punjab 140001, India
krajiv@iitrpr.ac.in

Abstract. The present study investigates the effect of various heat treatments on the microstructure, mechanical properties, and electrochemical behavior of Ti-6Al-4V alloy. The heat treatment (HT) comprises of treating above and below beta transus temperature (T_β) followed by aging at 650 °C for 2 h and furnace cooling. Results reveal that heat treatment temperature and cooling rate significantly modify the microstructure, mechanical, and electrochemical behaviour. Notably, the beta quenched and aged sample (HT-1) exhibited the highest tensile strength (1166 MPa) possibly due to the formation of α' martensite. On the other hand, the microstructure of the beta annealed and aged sample (HT-2) consists of Widmanstatten α/β phase. Conversely, the microstructure of the solution treated sample (HT-3) contains globular primary α_P and acicular α', which resulted in balanced mechanical properties (tensile strength of 1124 MPa and 18% elongation). Duplex annealed sample (HT-4) exhibits globular α_P and Widmanstatten α/β phase. Furthermore, it is seen that the best corrosion resistance is observed during HT-1 condition. Research findings show crucial insights into tailoring the microstructure, mechanical, and electrochemical properties of Ti-6Al-4V alloy through precise heat treatment.

Keywords: Ti-6Al-4V alloy · Heat treatment · Microstructure · Mechanical properties · Corrosion behavior

1 Introduction

Ti-6Al-4V alloy belongs to a family of titanium alloys having two phases, α and β, in its microstructure. The alloy combines a high strength-to-weight ratio, good toughness, and outstanding corrosion resistance [1, 2]. It is widely used in the manufacturing of

A. Tiwari et al. (Eds.): iConFIRM 2023, SPPHY 308, pp. 59–70, 2024.
https://doi.org/10.1007/978-981-97-4557-9_7

engineering components in different sectors such as aerospace, marine, medical, chemical, and automotive industries [3, 4]. Literature shows that this alloy contributes more than 50% of all titanium usage.

Researchers have shown that microstructure affects mechanical as well as electrochemical behavior. The microstructure of Ti-6Al-4V alloy can be controlled through different heat treatments and processing routes. For example, heating in the upper $\alpha + \beta$ or near the β-transus temperature region, followed by water quenching (WQ) increases tensile strength and hardness at the expense of elongation because of the formation of the brittle α' martensite phase [5]. On the other hand, when heated at a lower temperature in $\alpha + \beta$ region followed by the furnace or air cool (AC), it enhances the percentage elongation with reduction in tensile strength. R. Sahoo et al. [6] achieved UTS of 915–943 MPa with 18–23% elongation. Achieving UTS of above 1100 MPa with ductility more than 20% has been the challenge in wrought Ti-6Al-4V alloy. In addition, Ti-6Al-4V alloys are also employed in corrosive environments such as marine and gas turbines. The corrosion resistance of Ti-6Al-4V alloy is a critical fundamental behavior that should be considered for practical use. Titanium and its alloy have good corrosion resistance, because titanium forms a stable TiO_2 oxide film on its surface. W.Q. Lu et al. [7] reported the role of Al_2O_3 and V_2O_5, along with TiO_2, in forming a passive film. The objective of this research is to study the effect of different heat treatments on the microstructure, the mechanical properties, and the corrosion behavior of Ti-6Al-4V alloy to suit different engineering needs.

2 Experimental Details

The Ti-6Al-4V alloy was received in the form of plate of dimension 350 mm × 290 mm × 13 mm. The composition of the alloy is given in Table 1.

Table 1. Chemical composition (wt. %) of Ti-6Al-4V alloy used in this study

Material Condition	Al	V	Fe	C	O	N	H	Ti
Ti-6Al-4V alloy	6.39	4.21	0.19	0.034	0.113	0.028	0.0028	Bal

Samples with dimension 120 mm × 13 mm × 8 mm were used for heat treatment (HT). Titanium alloy has very high affinity towards oxygen. In order to eliminate the effect of oxidation, approximately 2 mm thick layer were removed from all sides of the HT samples. In this study, four different kinds of heat treatment, viz. HT-1, HT-2, HT-3, and HT-4 were carried out. The β transus (T_β) for Ti-6Al-4V alloy is 995 °C [6]. The details of HT temperature, time, and cooling medium specific to these HTs in this study are given below,

HT-1: 1050 °C for 0.5 h followed by WQ and then aging at 650 °C for 2 h
HT-2: 1050 °C for 0.5 h followed by AC and then aging at 650 °C for 2 h
HT-3: 950 °C for 0.5 h followed by WQ and then aging at 650 °C for 2 h, and
HT-4: 950 °C for 0.5 h, followed by AC and then aging at 650 °C for 2 h.

These heat treatments are also shown schematically in Fig. 1. All the HTs were carried out in Metrex microprocessor programmable muffle furnace having a temperature control of ±2 °C. Tensile test specimens were fabricated by use of wire-cut electric discharge machining (EDM) as per ASTM E8 standard. Tensile tests were carried out using Instron-8862 servo hydraulic machine at room temperature and strain rate of 10^{-3} s^{-1}. Vickers hardness were measured using an Economet BRVH-250DX tester.

Conventional metallographic techniques were used to prepare the samples for the microstructural analysis. The samples were etched with Kroll's reagent (0.5 ml HF, 1.5 ml HNO_3, and 18 ml H_2O) for 7 to 14 s. Microstructures were examined using Zeiss Vert. A1 inverted optical microscope and a Nova Nano FE-SEM 450 (FEI) scanning electron microscope. X-ray diffraction (XRD) analysis was performed using a Panalytical X'Pert Pro diffractometer. In all XRD studies, the incident radiation of Cu Kα (1.54 Å) within the angular range of (30–90) ° was used. The phase identification was done using X'Pert HighScore Plus software. Electrochemical studies were carried out in 3.5 wt.% NaCl solution using K-Lyte Z and PG Lyte 1.0 workstations. EC-Lab V11.30 and ZsimDemo 3.20d softwares were used to analyse the Tafel plots and electrochemical impedance spectroscopy (EIS) results respectively.

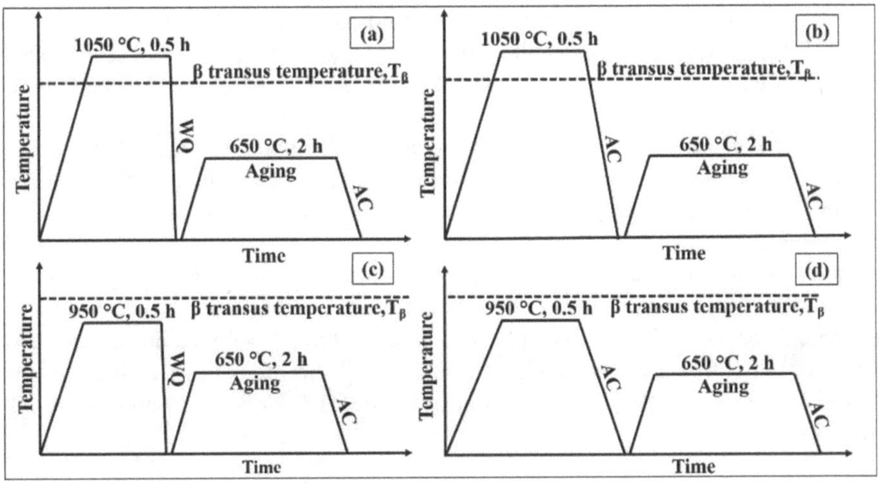

Fig. 1. Schematic diagram of different heat treatments (a) HT-1, (b) HT-2, (c) HT-3, and (d) HT-4.

The specimen was utilised as a working electrode, a standard calomel electrode (SCE) was used as a reference, and a platinum electrode was employed as a counter electrode in a three-electrode configuration. Potentiodynamic polarisation was observed at a scan rate of 1 mVs^{-1}. At the corrosion potential, EIS tests were carried out from 10^{-2} to 10^5 Hz with an amplitude of 10 mV.

3 Results and Discussion

3.1 Microstructural Characterization

The microstructure of heat treated Ti-6Al-4V alloy is shown in Fig. 2. As seen in Fig. 2(a, b), water quenching from pure β region and subsequent aging (HT-1), causes the transformation of pure β to majority of acicular α' martensitic microstructure with nano β particles (formed during second stage of aging at 650 °C for 2 h). S. A. Etesami et al. [4] carried out detailed study and reported the presence of nano β particles during aging at 611 °C for 2 h. However, transformation of the β phase into basketweave or Widmanstatten structure was observed in HT2, as shown in Fig. 2(c, d). The prior β grain has been seen to have a consistent polygonal shape (Fig. 2c).

The heat treatment results in the formation of plates consisting of α and β phases, which can be clearly observed in Fig. 2(d). The microstructure obtained during heat treatment HT-3 consists of primary alpha (α_p) and α' phase as shown in Fig. 2(e, f). M. Paghandeh et al. [8] observed a similar dual phase microstructure (α + α'). Figure 2(g, h) shows the microstructure produced during HT-4, which consists of primary α_p and lamellar α/β.

In order to identify the different phases present, XRD study of different heat-treated samples (HT1, HT2, HT3, HT4) have been carried out (Fig. 3). The XRD pattern taken during HT-1 shows peaks corresponding to α' and β. The α' (diffusionless transformation) is observed due to fast cooling. A similar observation is reported by Yu-Wei Cui et al. [9].

Both the α' martensite and α-phases possess a hexagonal close-packed (hcp) crystal structure, with similar lattice features. Consequently, distinguishing between the two becomes challenging. However, Martensitic α' is a solid solution of vanadium that is supersaturated. The diffusion-less transformation because of water quenching does not allow the diffusion of vanadium in α' phase compared to α, leading the supersaturation or higher vanadium content leads to more distortion in the lattice structure [10, 11]. The β peak is observed in HT-1 sample, which might have developed during the second stage of heat treatment (refer to Fig. 1), suggesting the breakdown of martensitic α' [12]. Intensity peaks of β phase was not detected in HT3. However, α and β phases are observed in HT-4.

3.2 Mechanical Properties

The engineering stress-strain curves of different samples (HT-1, HT-2, HT-3, and HT-4) are displayed in Fig. 4a. The yield strength (0.2%YS), ultimate tensile strength (UTS), elongation, (%El.) obtained during pertinent heat treatment is shown in Fig. 4b. It can be seen that the maximum YS (1102 MPa) and UTS (1166 MPa) are observed for HT-1 sample. However, ductility was minimum (8%). The higher strength is attributed to the presence of α' in the microstructure (Fig. 2a–b) which formed during water quenching from 1050 °C. The presence of α' reduced the ductility. The brittle nature of α' has also been reported [5].

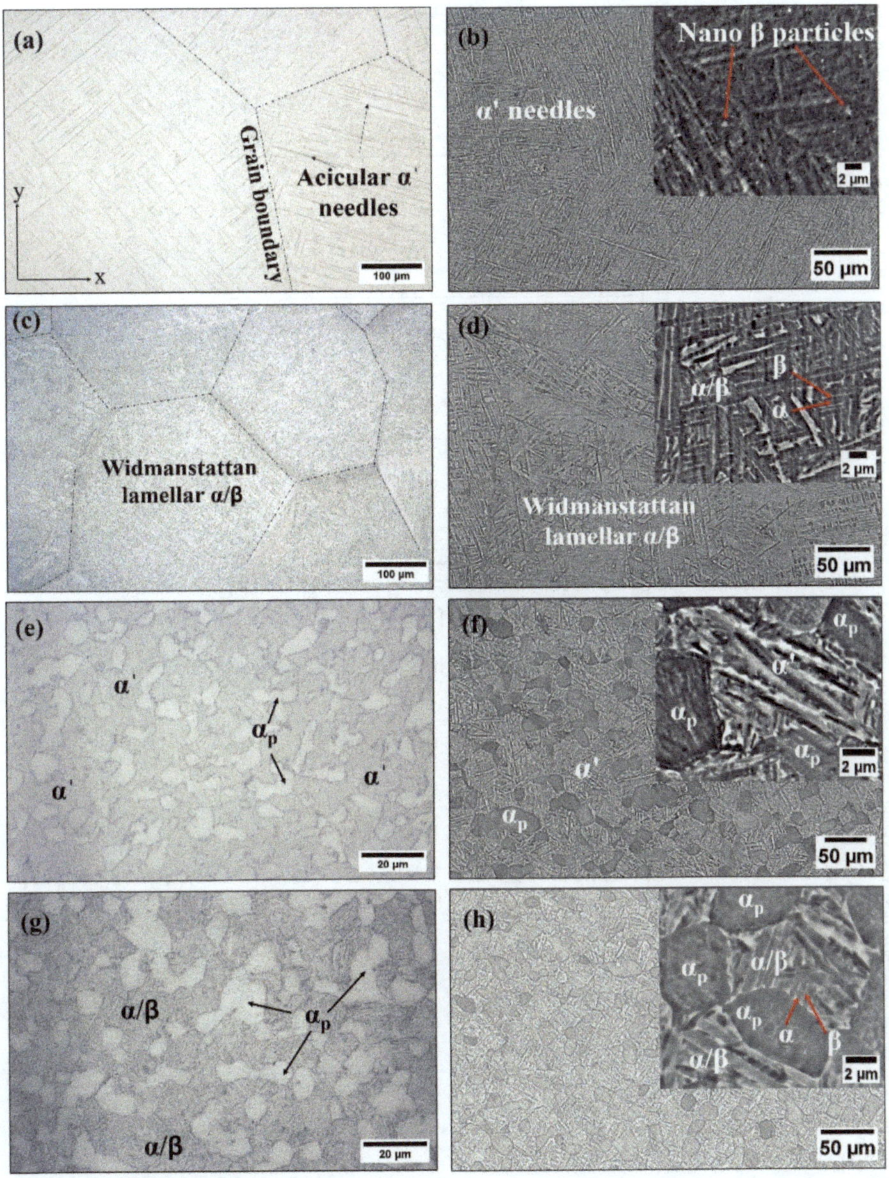

Fig. 2. Optical (left column) and SEM (right column) images of heat-treated samples: (a, b) HT-1, (c,d) HT-2, (e, f) HT-3, and (g, h) HT-4.

The ductility of HT-2 is better than HT-1 due to lamellar $\alpha + \beta$. An excellent combination of strength and ductility is found in HT-3 due to α_p and martensitic α'. HT-4 samples displayed lower strength than HT-3 and slightly higher ductility because of combination of α_p and lamellar $\alpha + \beta$ phases.

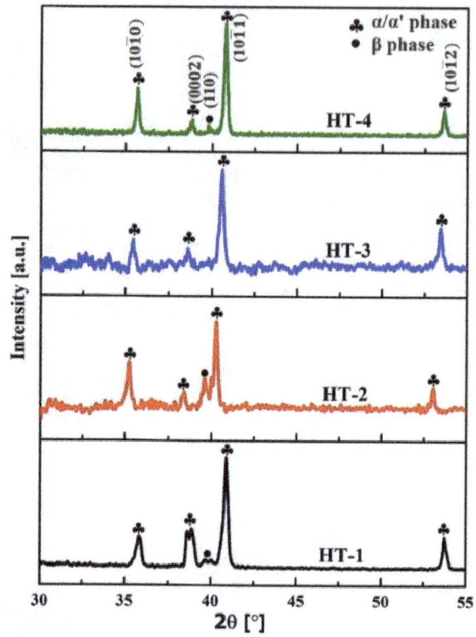

Fig. 3. XRD diffractograms of different heat-treated specimens.

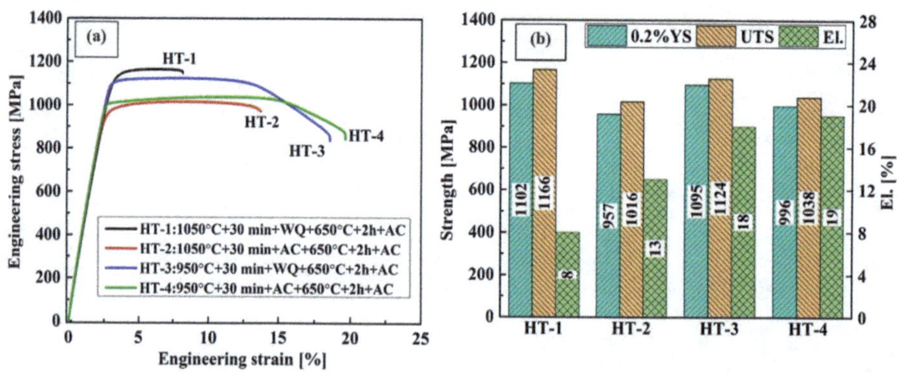

Fig. 4. Tensile results, (a) engineering stress-strain curves, (b) comparison of strength and ductility of four heat-treated Ti-6Al-4V alloy.

The Vickers hardness of HT-1, HT-2, HT-3 and HT-4 were found to be 374 ± 10, 360 ± 7, 373 ± 4, and 341 ± 6 respectively. A change in hardness value is observed under different HT condition. The Highest hardness observed in HT-1. This is consistent with tensile results where highest strength is obtained during HT-1 heat treated condition. Hardness value is dependent on microstructure developed during heat treatment. High hardness during HT-1 and HT-3 condition is ascribed to the presence of α'.

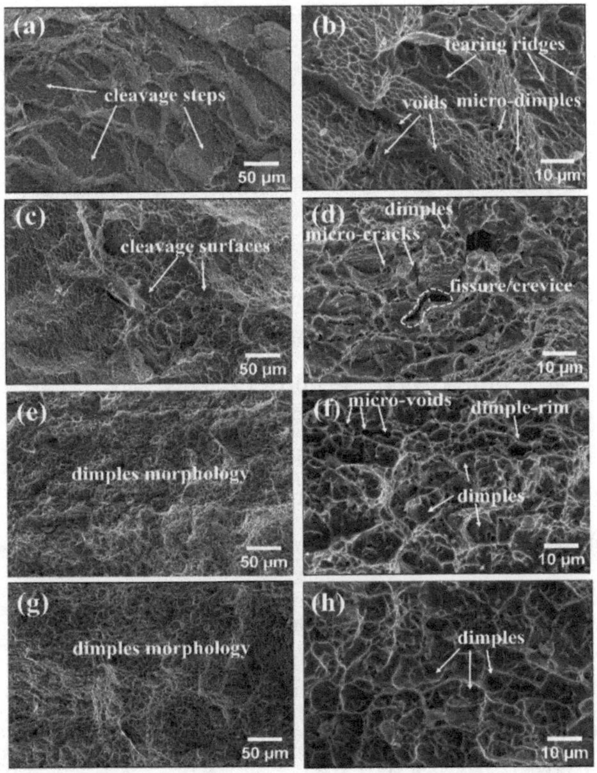

Fig. 5. SEM images of fracture surfaces of heat-treated samples after tension test: (a, b) HT-1, (c, d) HT-2, (e, f) HT-3, and (g, h) HT-4 at (1000X) and (5000X) respectively.

Fractographs of heat treated tensile tested samples are shown in Fig. 5. The prominent features (Fig. 5(a, b)) of HT-1 are facets, ripping ridges, and few threaded dimples. The fractographs indicate cleavage fracture mechanism. The presence of fissures, micro-cracks, and dimples shown in Fig. 5(c, d) indicate mixed failure in HT-2 with improved ductility. In contrast to HT-1 (Fig. 5(a, b) and HT-2 (Fig. 5(c, d) dimpled fracture mechanics dominates during HT-3 (Fig. 5(e, f)) and HT-4 (Fig. 5(g, h)). HT-3 and HT-4 exhibit increased ductility due to the high fraction of α_p (Fig. 2(e, g)) [13].

3.3 Electrochemical Analysis

All samples reach stable potential by dipping in solution within an hour. The measured stable potential for HT-1, HT-2, HT-3, and HT-4 were −0.33, −0.19, −0.198, and −0.15 V respectively.

Potentiodynamic Polarization. The values of corrosion potential (E_{corr}), corrosion current density (I_{corr}), passivation current density (I_{pass}), and corrosion rate (CR) are determined from tafel curves as shown in Fig. 6, and its values are given in Table 2. The measured curve shows a similar pattern for all samples.

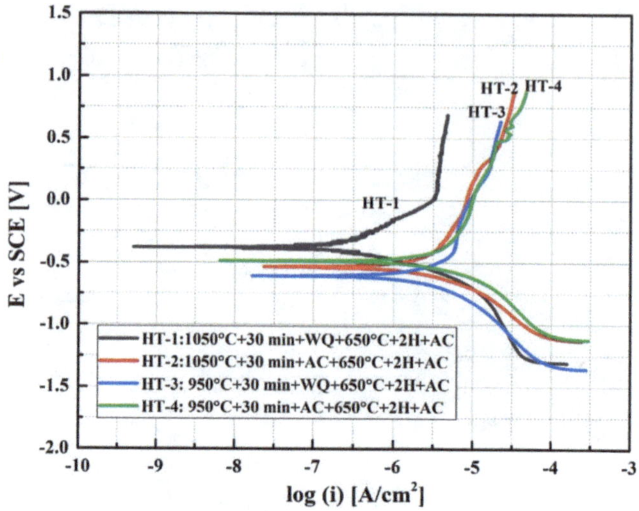

Fig. 6. Potentiodynamic polarization curves of the samples in 3.5 wt.% NaCl solution.

Table 2. Electrochemical parameters of heat-treated specimens in 3.5 wt.% NaCl solution

Ti-6Al-4V	Ecorr (V)	Icorr (μA/cm^2)	Ip (μA/cm^2)	CR (μpycm^{-2})
HT-1	-0.37 ± 0.02	0.21 ± 0.04	4.84	1.88
HT-2	-0.60 ± 0.01	1.61 ± 0.42	9.08	14.27
HT-3	-0.61 ± 0.04	1.86 ± 0.42	22.02	16.43
HT-4	-0.64 ± 0.08	2.63 ± 0.86	46.57	23.29

Passivation begins when the potentiodynamic curve deviates from the Tafel anodic line when corrosion potential (E_{corr}) of HT-1 exceeds 0 V. NaCl solution combines with Ti-6Al-4V alloy to form a dominating TiO_2 passivation film (Eqs. (1)–(5)) [7, 14].

$$Ti^{4+} + 6Cl^- \rightarrow TiCl_6^{2-} \quad (1)$$

$$2Ti + 3H_2O \rightarrow Ti_2O_3 + 6H^+ + 6e^- \quad (2)$$

$$Ti + 3H_2O \rightarrow TiO(OH)_2 + 4H^+ + 4e^- \quad (3)$$

$$Ti_2O_3 + 3H_2O \rightarrow 2TiO(OH)_2 + 2H^+ + 2e^- \quad (4)$$

$$TiO(OH)_2 \rightarrow TiO_2 + H_2O \quad (5)$$

HT-1 showed lowest CR suggesting the best corrosion resistance. Lowest passivation current density was observed for HT-1, indicating easy formation of passivation layer [15]. HT-2 and HT-3 had similar corrosion characteristics, indicating near equal corrosion properties. HT-4 has the lowest corrosion resistance.

Electrochemical Impedance Spectroscopy (EIS) Measurements. In order to study the characteristics of passive film formed, EIS measurements were carried out. Nyquist plots of different heat treated sample are presented in Fig. 7(a). HT-1,2 Nyquist impedance plot reveal incomplete semicircle arcs due to charge transfer at the electrolyte/electrode contact.

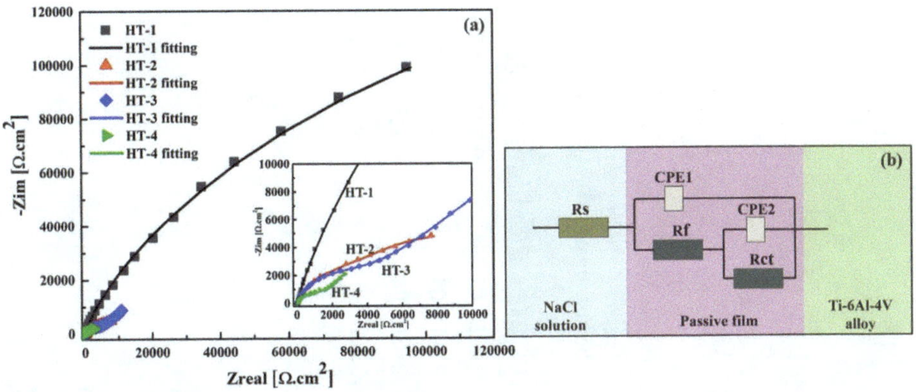

Fig. 7. (a) Nyquist plots of heat-treated Ti-6Al-4V alloys: HT-1, HT-2, HT-3 and HT-4 in NaCl solution (b) equivalent circuit.

It is known that corrosion resistance is related to semicircular arc diameter. A larger arc radius indicates higher corrosion resistance. The imperfect semicircle diameter of HT-1 sample is larger indicating better passivation behavior than HT-2, HT-3, and HT-4 samples.

The equivalent circuit, shown in Fig. 7(b), reasonably fits the experimental results obtained from EIS measurements. The corresponding fitting parameters are given in Table 3. R_s, R_f, and R_{ct} are the resistance of solution, passive film, and charge transfer resistance of bulk material respectively. CPE_1 and CPE_2 are the capacitances of passive film and electrochemical double layer respectively. Fit parameter goodness is indicated by χ^2 values. Fitting improves with decreasing values. The n_1 and n_2 values show the constant phase element (CPE) from ideal capacitor to ideal resistor when changed from 1 to 0.

It can be seen from Table 3 that the value of R_{total} is highest for HT-1 indicating highest corrosion resistance.

Table 3. EIS test parameters of the heat-treated Ti-6Al-4V alloy

Ti-6Al-4V	R_s (Ω.cm^2)	CPE1 (S.sec^n/cm^2)	n_1	R_f (Ω.cm^2)	CPE2 (S.sec^n/cm^2)	n_2	R_{ct} (Ω.cm^2)	χ^2	R_{total} (Ω.cm^2)
HT-1	31.18	2.76E−05	0.927	1318	1.82E−05	0.492	6.85E + 05	8.32E−03	6.86E + 05
HT-2	27.74	4.04E−05	0.875	5719	2.30E−04	0.667	3.24E + 04	5.59E−03	3.81E + 04
HT-3	29.37	3.09E−05	0.894	3674	2.05E−04	0.524	3.50E + 04	9.68E−03	3.87E + 04
HT-4	30.27	2.74E−05	0.904	1240	3.80E−04	0.593	3.42E + 04	6.28E−03	3.55E + 04

4 Conclusion

In this study, Ti-6Al-4V alloy was given different heat treatment and the effect of microstructure evolution on mechanical and electrochemical behavior were analyzed. The following conclusions were drawn.

- It is observed that heat treatment temperature and cooling affect the microstructure evolution which in turn modify the mechanical response of the alloy. The highest yield and tensile strength were observed 1166 MPa and 1102 MPa respectively during HT1.
- The best customized mechanical properties (UTS, 1124 MPa; ductility 18%) is obtained during HT-3. The optimum strength and ductility obtained during HT-3 is attributed to the formation of acicular α' and globular α_P.
- Potentiodynamic polarization and EIS results showed highest corrosion resistance during HT-1 condition.

References

1. Lutjering, G., Williams, J.C.: Titanium, 2nd edn. Springer, Berlin (2007)
2. Schuch, M., Hahn, T., Herzig, N., Bleckmann, M.: The mechanical behavior and microstructure of additively manufactured Ti6AI4V for different material states and loading conditions. J. Dyn. Behav. Mater. **8**, 273–284 (2022)
3. Donachie, M.J.: Titanium A Technical Guide, 2nd edn. ASM International, USA (2000)
4. Etesami, S.A., Fotovvati, B., Asadi, E.: Heat treatment of Ti-6Al-4V alloy manufactured by laser-based powder-bed fusion: process, microstructures, and mechanical properties correlations. J. Alloys Compd. **895**, 162618 (2022)
5. Reda, R., Nofal, A., Hussein, A.H.: Effect of single and duplex stage heat treatment on the microstructure and mechanical properties of cast Ti-6Al-4V alloy. Metallogr. Microstruct. Anal. **2**, 388–393 (2013)
6. Sahoo, R., Jha, B.B., Sahoo, T.K.: Effect of primary alpha phase variation on mechanical behaviour of Ti-6Al-4V alloy. Mater. Sci. Technol. **31**, 1486–1494 (2015)
7. Lu, W.Q., Liu, Y.J., Wu, X., Liu, X.C., Wang, J.C.: Corrosion and passivation behavior of Ti-6Al-4V surfaces treated with high-energy pulsed laser: a comparative study of cast and 3D-printed specimens in a NaCl solution. Surf. Coatings Technol. **470**, 129849 (2023)
8. Paghandeh, M., Zarei-Hanzaki, A., Abedi, H.R., Vahidshad, Y.: On the warm temperature strain accommodation mechanisms of Ti–6Al–4V alloy holding different starting microstructures. J. Mater. Res. Technol. **14**, 496–506 (2021)
9. Cui, Y.W., et al.: Metastable pitting corrosion behavior and characteristics of passive film of laser powder bed fusion produced Ti–6Al–4V in NaCl solutions with different concentrations. Corros. Sci. **215**, 111017 (2023)
10. Ettefagh, A.H., Zeng, C., Guo, S., Raush, J.: Corrosion behavior of additively manufactured Ti-6Al-4V parts and the effect of post annealing. Addit. Manuf. **28**, 252–258 (2019)
11. Paghandeh, M., Zarei-Hanzaki, A., Abedi, H.R., Vahidshad, Y., Minarik, P.: The correlation of c-to-a axial ratio and slip activity of martensite including microstructures during thermomechanical processing of Ti–6Al–4V alloy. J. Mater. Res. Technol. **18**, 577–583 (2022)
12. Kaschel, F.R., et al.: Mechanism of stress relaxation and phase transformation in additively manufactured Ti-6Al-4V via in situ high temperature XRD and TEM analyses. Acta Mater. **188**, 720–732 (2020)

13. Zhao, Z., Chen, J., Tan, H., Zhang, G., Lin, X., Huang, W.: Achieving superior ductility for laser solid formed extra low interstitial Ti-6Al-4V titanium alloy through equiaxial alpha microstructure. Scr. Mater. **146**, 187–191 (2018)
14. Pazhanivel, B., Sathiya, P., Muthuraman, K., Sozhan, G.: Influence of NaCl environment on stress corrosion cracking of additive manufactured Ti-6Al-4V alloy. Eng. Fail. Anal. **127**, 105515 (2021)
15. Ding, X., Ma, H., Zhang, Q., Yang, J., Li, D., Fan, S.: Effect of annealing heat treatment on microstructure and corrosion behavior of Ti6Al4V alloy fabricated by multi-laser beam wire-feed additive manufacturing in vacuum environment. J. Alloys Compd. **914**, 165363 (2022)

Finite Element Modelling in Bones: A Review

Kirti Singh[1] and Abhishek Tiwari[2(✉)]

[1] Department of Biomedical Engineering, Indian Institute of Technology (IIT) Ropar, Rupnagar 140001, India

[2] Fracture Mechanics Lab, Metallurgical and Materials Engineering, Indian Institute of Technology (IIT) Ropar, Rupnagar 140001, India

abhishek.tiwari@iitrpr.ac.in

Abstract. This review explores the utilization of Finite Element Method (FEM) modelling in the field of bone biomechanics. It emphasizes the crucial significance of bone biomechanics within the field of medical research for therapeutic applications. The versatility of finite element method (FEM) is demonstrated in its application to biomechanical investigations, clinical applications, and disease modelling. Nevertheless, this study also acknowledges and discusses the obstacles and potential avenues for further exploration within the discipline. The discussion revolves around the careful management of model complexity and computing efficiency, the need for accurate material characterization, the importance of interdisciplinary collaboration, and the ethical considerations involved. In its whole, the review illustrates the significant role of finite element modelling (FEM) in advancement of bone biomechanics research and its practical applications.

Keywords: FEM · bone biomechanics · numerical methods · UMAT · Abaqus · medical research

1 Introduction

The word "biomechanics" refers to the application of mechanical laws to biological systems in order to comprehend how the body moves. Biological systems, in contrast to other bodies, are always changing, which frequently causes issues when attempting to apply generic mechanical concepts to such living systems [1]. Gaining a better knowledge of how the bone adapts to mechanical pressures for bone growth or fracture repair is the focus of bone biomechanics [2]. In order to develop useful treatments for a variety of disorders, medical research and engineering rely heavily on this understanding of the quality and fragility of bone, which is important in the progression of many diseases such osteoporosis [3, 4]. The mechanical behavior of the skeletal system has been studied using finite element modelling (FEM) since 1972. It offers a means of analyzing the whole stress field and representing the overall irregular bone structure [5]. In order to create treatment or rehabilitation modalities for the patient, FEM helps to understand how the disease and fracture progresses in the bone in response to mechanical stimuli [6].

A. Tiwari et al. (Eds.): iConFIRM 2023, SPPHY 308, pp. 71–81, 2024.
https://doi.org/10.1007/978-981-97-4557-9_8

This review focuses on the biomechanical properties of bone and their utilization as an input for developing a finite element analysis (FEA) model. It aims to establish a framework for model creation and discuss the applications of such models in the field of medical research along with the challenges of utilizing FEA.

1.1 Bone Structure and Properties

Two hundred and six bones make up an adult human's structural framework, and 99% of the body's calcium is found in these bones [7]. The architecture of the bone is composed of a hierarchical arrangement of variously ordered components that work together to carry out biochemical and mechanical tasks. Collagen fibers that have been mineralized and stacked in sheets to produce lamellae make up the tiny bone structure. Osteon is formed by the concentric arrangement of these lamellae around the Haversian canal. On the other hand, a tangential arrangement of lamellae generating a stack of layers with woven bone forming the lamellar bone and 50–300 μm rod-like interconnecting trabecular bone, which makes up the cancellous bone, is observed in some bones without osteons [8].

Cortical Bone

It is a transversely isotropic substance with a porosity of 5% to 15%, with variations in porosity accounting for 75% of the bone's strength variation. The cortical bone's brittleness increases at high strain rates, with greater brittleness in the transverse direction than the longitudinal, and loading rate is a major factor in the production of microcracks in the bone. With a sudden fracture strain of less than 3%, the tension ceases. Unlike compressive longitudinal loading, which shows the maximum strength during longitudinal compressive loading, it displays a bilinear stress-strain response during tension, with simultaneous hardening and softening occurring before failure at 1.5% strain.

Cancellous Bone

It has a high porosity anisotropic substance and is also referred to as trabecular bone. Its strength is greatest during compression and lowest in shear, similar to cortical bone. At 50% compressive strains, the spongy bone retains its load-bearing capability and allows about 0.7% compression. When the cancellous bone is compressed past its yield point (up to 5%), it loses its rigidity and strength, which illustrates the possibility of fractures brought on by progressive, irreversible deformations [9].

To attain such success rates, osseointegration (OI) is essential. In order to create a structural and functional connection between the implant surface and the living bone without progressive micromotion, it refers to mechanical and biological anchoring [10–12]. Loss of bone-to-implant contact (BIC) could jeopardise osseointegration if there is a persistent disruption in the equilibrium between bone resorption and apposition (Bosshardt et al., 2016). Micromotion greater than 150μm, which results in low OI and reduced primary and secondary stability, can lead to unsuccessful healing [13]. Bone healing is comprised of three partially overlapping steps:

1. Inflammation Phase
2. Repair Phase

3. Remodeling Phase

The healing process begins with the inflammatory phase and can continue for up to seven days. Polymorphonuclear neutrophils (PMNs) are released in reaction to tissue damage caused through the production of pro-inflammatory chemicals such as cytokines, growth factors, and chemokines [14]. Because hematoma formation occurs during this phase, which is essential for the effective settlement of fractures, it is sometimes referred to as the hematoma phase [15]. Following the clearance of the hematoma, endochondral ossification—a cartilaginous callus development—occurs, culminating in the production of a hard callus osseous tissue on the periphery of the soft callus region [15, 16]. Finally, bone remodelling is the outcome of mechanobiological controls. This dynamic physiological process includes bone apposition and resorption [15, 17]. It starts with the activation and resorption of bone by osteoclasts, which is then followed by the mineralization of the newly deposited organic bone matrix by osteoblasts [18]. Numerous mechanical forces propel the process of remodelling. Mechanical pressures serve as primary cues that, through mechanosensitive signalling pathways, are converted into biochemical signals that drive bone formation and healing. This process is known as bone adaptation to the environment [19, 20].

Along with a number of other variables like loading parameters, material qualities, the quantity and quality of the BIC, the quality and quantity of the surrounding bone, and so on, the surface topography and design features determine the implant-bone response, particularly during the bone remodelling phase [21, 22]. When implants with a limited radius of curvature or geometric discontinuities were subjected to vertical loads, Siegele et al. demonstrated notable changes in the way that stress was dissipated in the bone. Additionally, they showed that during lateral loading, the soft tissue layer and direct implant-bone contact area experienced the highest concentration of stress in the crestal region for screw implants. The use of oblique loads for FE analysis was recommended by Holmgren et al. in another work, claiming that they replicate accurate occlusion directions that produce the highest localized stresses in the cortical bone [23].

2 Numerical Studies

A 3D part of the skeletal system is generated from the computed tomography (CT) scan or three-dimensional (3D) scan post extraction of the external curvature from obtained slices using an algorithm for border tracing. The extracted geometry is meshed using softwares with an automatic meshing routine such as Hypermesh. A recent development in automatic mesh generation (AMG) utilising voxel meshes instead of hexahedral or tetrahedral meshes as it allows the user to bypass the geometry extraction procedure [26, 27]. Moreover, a study conducted by Austman et al. demonstrated comparatively better accuracy when a second-order tetrahedral mesh was used instead of triangular surface meshes [28]. Kluess et al. reported use of hexahedral mesh for better accuracy and faster computational time in case of contact definition problems [29]. It is pertinent to perform a mesh sensitivity analysis for model optimization after the assignment of the relevant material properties and interactions. Bone is an anisotropic material but has been modelled as an isotropic or transversely isotropic material as well. For instance, a study carried out on the femur bone of humans assumed isotropic material properties for bone due to the tedious process of assigning the entire bone anisotropic properties [30]. While developing a model for simulation of the human body parts, it is necessary to take into consideration the various human factors which vary in a population and individual. Hence, Eq. (1) describes the relationship between the Young's modulus, strain rate and bone density to simulate bones of different quality [31].

$$E = 3790 \times \rho \times^{3} \varepsilon^{0.06} \tag{1}$$

A simulation expert often needs to develop a simplistic model to minimize the model running times but also take into account that the developed model is a close approximation of the real-world scenario. This involves applying boundary conditions in consideration with the assumptions made for the model. Nonetheless, it is to be noted that the forces, pressure or applied displacements should mimic the physiological conditions [32].

Recent developments in the field of medical devices have prompted varied studies to simulate the dynamic modelling and remodelling process of bone using VUMAT or UMAT subroutines. Humbli and Thurner expressed the bone remodelling process using a Continuum Damage Mechanics (CDM) approach demonstrated in Eq. (2).

$$\sigma_{ij} = \left(1 - D_{fat}\right) \times a_{ijkl} \times \varepsilon_{kl} \tag{2}$$

where,

$$D_{fat} = 1 - [1 - \left(\frac{N}{N_f}\right)^{\frac{1}{1-\gamma}}]^{\frac{1}{1+\beta}} \tag{3}$$

Here, and are material parameters and N_f is the cycle of failure for compressive or tensile loading. Further, the Young's modulus for the mineralized or damaged bone can be calculated as,

$$E = C \times \left(1 - D_{fat}\right) \times \rho^{p} \times \alpha^{q} \tag{4}$$

where, ρ is the calculated density using Komarova's model and α is the ash function denoting mineralization of bone. C is experimentally derived and according to Hernandez et al. $q = 2.74$ meanwhile $3 > p > 2$.

[33] Weinans et al., describes the rate of change of bone density as a function of mechanical stimulus S and the threshold stimulus k as described in Eq. (5).

$$\frac{d\rho}{dt} = B \times (S - k) \quad \rho_{cb} \geq \rho > 0 \tag{5}$$

where, B represents a constant and ρ_{cb} the maximum density of bone. Furthermore, if the threshold value (k) is greater than the loading, resorption will occur; conversely, if the loading is greater than 'k', bone growth will take place. However, Weinan's model fails to address the overload resorption occurring due to large loadings.

Validation and verification of the model through experimentation is crucial for model calibration. Statistical analysis such as analysis of variance (ANOVA) should be performed to validate the FEA model whereby, the standard deviation (SD) between the results from the simulation analysis and experimentally analyzed average data shouldn't exceed 1 with an >0.05 to reject the null's hypothesis in ANOVA [34, 35].

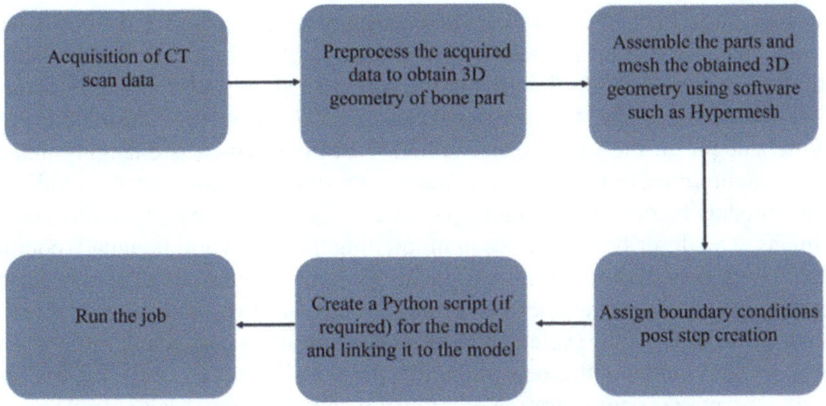

Fig. 1. Schematic representation of the process involved in model creation.

3 Applications

Advancements in healthcare have led to an increase in awareness about movement disorders which often result in death or fracture due to inappropriate gait. Gait disorders can stem from varied clinical disorders such as osteoarthritis, Parkinson's, stroke or due to old age [36]. FEA is used to predict loading conditions while standing or moving.in the parts of the foot which is useful for improvement of gait and development of rehabilitation devices [37]. Additionally, industries and research organizations perform FEA for developing and optimizing implant designs for various parts of the body. For instance, Ferdinand et al. developed an anterior-cruciate ligament (ACL) compatible design for knee implant using a Mooney-Rivlin model with transverse isotropy and X, Y translation and Z rotation connectors for the knee joint motion. This led to the development of an implant design for bi-cruciate knee arthroplasty [37]. FEA has also been used for management of fracture fixation sometimes utilizing dynamic solver with complex contacts and subroutines predicting fracture healing or progression of diseases such as osteoporosis resulting in bone porosity change and efficiency of the implant for fracture fixation [38, 39].

4 Challenges and Future Directions

Currently, there have been very few studies incorporating the piezoelectricity induced mechanoadaptation in bone with Bansod et al. performing the first study utilizing a python script for femur bone remodeling with piezoelectric effect. However, hexahedral meshing is not embedded in the software used for this study so, a ten-noded tetrahedral mesh was utilized for the domains and the study was performed for 100 days with a step time interval (t) of 0.1 day. However, experimental validation of the simulated data in case of human bones is not always possible due to the unavailability of patient specific human models [40] resulting in hampered accuracy [6]. Additionally, bone remodeling is a complex process involving complex theoretical equations which are sometimes oversimplified for simulations thereby, not remaining relevant in the clinical practice.

Prior utilization of FEA in dentistry involved simplifying parameters such as boundary conditions, material properties, geometry, and load [24, 25]. In future, these models can be optimized to improve the validation of the bony structures and studying implant fixation [41]. Furthermore, characterization of bone material properties is a tedious process largely dependent on the trabeculae orientation in bone which is changing and challenging to incorporate in a model [42] thereby, there is a requirement for studies with improved models incorporating algorithms to closely model the complexities of bone. Brekelmans et al. describe difficulties in mimicking the real-world boundary conditions associated with bone biomechanics and identifying the appropriate data to be incorporated into the model; however, they performed studies on a two-dimensional (2D) model instead of a 3D model of the skeletal part [43]. A recent study carried out by Cronin et al. described a CDM model with stress triaxiality, previously ignored in myriad studies due to lack of experiments demonstrating relationship between stress triaxiality. They developed an orthotropic model with the primary material axis in the direction of osteons and the remaining axis as radial and circumferential. Nonetheless, experimental verification to understand the reliability of such models still needs to be validated [44, 45].

Navigating these challenges and pursuing the future directions can lead to more accurate, reliable, and clinically applicable bone FEM models, ultimately advancing our understanding of bone biomechanics and its implications for healthcare and medical interventions.

5 Case Studies

A study conducted Mughal et al. to understand the reaction forces on the femur bone of humans. They developed a CAD model using the 3D scanning technique and performed post-processing involving mesh refinement to minimize unrealistic results. An average Young's modulus of 15 MPa was chosen for this analysis. Two independent simulations involving application of axial and perpendicular forces on the human femur bone were performed. They found the femur bone fractured when subjected to 414 kg and 60 kg weight in case of axial and perpendicular loading respectively. Hence, the study reported the superior performance of the bone under axial forces and indicated the criteria for probable femur bone fractures [29]. Additionally, it highlights the significance of FEA in comprehending the mechanical behavior of bone and it's performance under different loading conditions elucidating the crucial role FEA plays in orthopedic biomechanics and implant design.

To understand the behaviour of tibia subjected to torque, Carbonell et al. performed a study to demonstrate patient-specific bone remodeling whereby CT images of the bone were obtained in digital imaging and communication in medicine (DICOM) format and reconstructed using MIMICS software. Post, these mechanical properties were ascribed to the obtained 3D geometry after importing the file in Abaqus 6.13 (SIMULIA, Dassaults Systems, RI, USA). The complex geometry was meshed using C3D4 tetrahedral mesh creating an unstructured mesh with an average shape factor of 0.656. Assuming a bone remodeling equilibrium, the used stress was defined using Eq. (6), whereby U and E describe the strain energy and elastic modulus respectively for continuum stress.

$$\sigma_{\varepsilon} = \sqrt{2EU} \tag{6}$$

Furthermore, the bone density change was described as shown in Eq. (7). It represents remodeled bone tissue density and k is the actively remodeling local area fraction. Martin quantified the actively available surface area bone volume for remodeling using S_V.

$$\Delta\rho = k \cdot S_V \cdot \rho_t \quad (7)$$

The algorithm was further written using UMAT subroutine in FORTRAN to simulate bone remodeling process after application of boundary conditions on the model. During analysis, each iteration represented a day with a bone density of .5 g/cm^3 during the first step and the job completion marked the evident visual of the newly remodeled bone structure with cortical bone density of tibia approximately 1.92 g/cm^3 with the maximum density located near to the distal epiphysis. The model took into consideration the difference in bone remodeling in homogeneous and non-homogeneous bone and developed a model that could be used by clinicians during orthopedic treatments [46, 47].

Ridha Hambli et al. conducted a study using a novel approach, termed the multiscale methodology integrating FEA and neural network computation, for simulating bone remodeling. In this research, the apparent bone Young's modulus was determined by the neural network using factors such as trabecular bone structure, tissue modulus, and bone volume fraction. The recorded rise in the overall bone's average elastic modulus, starting at 4,000 MPa initially, results from both the increase in average density and the accumulation of damage. Parallelly a simulation model mimicking same conditions was run to derive a correlation between the model using the novel approach and the classical methodology. The study showcased comparable results between the two experiments but, the multiscale methodology completed the job thousand times faster than the classical approach [48].

Another study performed by Guha et al. elucidates the crucial role of finite element analysis in understanding disease progression and treatment planning. The study involves utilizing CT scan data to closely capture the microstructure of trabecular bone. Furthermore, the mechanical behavior of the 3D model under different loading conditions is simulated using a continuum mechanical modelling approach. The study found that CT-derived trabecular bone (Tb) modulus values obtained through continuum finite element analysis (FEA) demonstrated strong linear correlations with both micro-CT-derived reference values and experimentally measured values. The linear correlation between computed modulus values and reference values using linear FEA modeling was comparable to that achieved with nonlinear modeling. About 39.4% of voxels in the nonlinear finite element analysis (FEA) at 0.5% strain were found to be in the elastic phase, while 60.6% moved into the plastic phase during simulation. These results suggest that, in comparison to their reference values, the nonlinear component of the plastic phase is responsible for the lower values of the nonlinear FEA modulus. It's also critical to remember that in vivo imaging's constrained signal-to-noise ratio and spatial resolution contribute a number of artefacts, such as noise, partial voluming, structural blurring, and loss of thinner trabeculae. Despite these difficulties, the continuum modelling approach appears to effectively reduce the impact of stress-flow along trabecular bone microstructure, especially when used in conjunction with in vivo imaging, which has its own set of limitations and artefacts. This is indicated by the high Pearson correlation between continuum FEA and both micro-CT and mechanical test-derived reference values. Additionally, high repeatability of CT scan reproducibility was observed for computed modulus values using both linear and nonlinear continuum FEA. The study also noted that high-stress regions coincided with Tb microstructure, as characterized by bone mineral density (BMD) values. Furthermore, distributions of von Mises stress over Tb microstructure and marrow regions were found to be significantly different. In conclusion, voxel-based continuum FEA provides surrogate measures of Tb modulus from CT imaging, eliminating the need for segmentation of Tb and marrow regions while considering bone distribution at the microstructural level. This relaxation of binary segmentation extends the application of FEA to assess mechanical properties of bone microstructure with relatively low-resolution imaging [49].

6 Conclusion

In bone biomechanics research, Finite Element Method (FEM) modelling is an essential tool that provides insights into bone behavior under a range of mechanical stressors. This review demonstrates its versatility in biomechanical investigations, clinical applications, and disease modelling, underscoring its importance in medical research and therapeutic applications. Notwithstanding its usefulness, it acknowledges the difficulties in controlling model complexity, guaranteeing precise material characterization, interdisciplinary cooperation, and ethical issues. The review emphasizes how FEM has been successfully applied to comprehend bone properties and structure, including cortical and cancellous bone characteristics. It also explores the intricacies of bone repairing procedures and the significance of osseointegration in implant success. There includes a discussion of numerical studies that use FEM, with a focus on the significance of mesh production, material property assignment, and boundary condition application. The most recent developments in simulating bone remodeling processes and experimentally validating FEA models are also discussed. The impact of FEM in healthcare is demonstrated by elucidating its numerous applications, such as fracture repair, implant design, and gait analysis. The review does, however, also point out difficulties, including the need to address bone material property characterization, improve model accuracy, and incorporate piezoelectricity-induced mechanoadaptation. In spite of these difficulties, the review comes to the conclusion that overcoming them and going in new directions can result in more precise and clinically useful bone FEM models, which will ultimately advance our knowledge of bone biomechanics and its consequences for medical treatments and healthcare.

References

1. Roesler, H.: The history of some fundamental concepts in bone biomechanics. J. Biomech. **20**(11–12), 1025–1034 (1987)
2. Roberts, W.E., Huja, S., Roberts, J.A.: Bone modeling: biomechanics, molecular mechanisms, and clinical perspectives. Seminar. Orthodont. **10**(2), 123–161 (2004)
3. Turner, C.H.: Biomechanics of bone: determinants of skeletal fragility and bone quality. Osteoporos. Int. **13**(2), 97–104 (2002)
4. Beaupied, H., Lespessailles, E., Benhamou, C.-L.: Evaluation of macrostructural bone biomechanics. Joint Bone Spine **74**(3), 233–239 (2007)

5. Keyak, J.H., Meagher, J.M., Skinner, H.B., Mote, C.D.: Automated three-dimensional finite element modelling of bone: a new method. J. Biomed. Eng. **12**(5), 389–397 (1990)
6. Parashar, S.K., Sharma, J.K.: A review on application of finite element modelling in bone biomechanics. Perspect. Sci. **8**, 696–698 (2016)
7. Brandi, M.L.: Microarchitecture, the key to bone quality. Rheumatology **48**(suppl. 4), iv3–iv8 (2009)
8. Rho, J.-Y., Kuhn-Spearing, L., Zioupos, P.: Mechanical properties and the hierarchical structure of Bone. Med. Eng. Phys. **20**(2), 92–102 (1998)
9. Elise, F., Morgan, G.U.: Bone mechanical properties in healthy and diseased states. Annu. Rev. Biomed. Eng. **20**(1), 119–143 (2018)
10. Mavrogenis, R.D.A.F.: Biology of implant osseointegration. J. Musculoskeletal Neuronal Interact. 61–71 (2009)
11. María, B., Guglielmotti, D.G.: Research on implants and osseointegration. Periodontology **2000**, 178–189 (2019)
12. Hamdan, S., Alghamdi, J.A.: The development and future of dental implants. Dent. Mater. J. 167–172 (2022)
13. Dieter, D., Bosshardt, V.C.: Osseointegration of titanium, titanium alloy and zirconia dental implants: current knowledge and open questions. Periodontology **2000**, 22–40 (2016)
14. Masahiro Maruyama, C.R.: Modulation of the inflammatory response and bone healing. Front. Endocrinol. (2020)
15. Mohammad, S., Ghiasi, J.C.: Bone fracture healing in mechanobiological modeling: a review of principles and methods. Bone Rep. **6**, 87–100 (2017)
16. Werner Winter, D.K.: Micromotion of dental implants: basic mechanical considerations. J. Med. Eng. (2013)
17. Gerstenfeld, T.A.: Fracture healing: mechanisms and interventions. Nat. Rev. Rheumatol. 45–54 (2015)
18. Bente Langdahl, S.F.: Bone modeling and remodeling: potential as therapeutic targets for the treatment of osteoporosis. Therap. Adv. Musculoskeletal Dis. 8, 225–235 (2016)
19. MS, J.S.-H.-R.: Encyclopedia of Nanotechnology. Springer, Dordrecht (2012). https://doi.org/10.1007/978-94-017-9780-1
20. Lijun Wang, X.Y.: Mechanical regulation of bone remodeling. Bone Res. (2022)
21. El-Anwar, M.I., M.M.-Z.-M.: Load transfer on dental implants and surrounding bones. Austral. J. Basic Sci. 551–560 (2012)
22. Samira Faegh, S.M.: Load transfer along the bone–dental implant interface. J. Biomechan. **43**(9), 1761–1770 (2010)
23. Dincer Bozkaya, S.M.: Evaluation of load transfer characteristics of five different implants in compact bone at different load levels by finite element analysis. J. Prosthetic Dentistry **92**(6), 523–530 (2004)
24. Trivedi, S.: Finite element analysis: a boon to dentistry. J. Oral Biol. Craniofacial Surg. **4**(3), 200–203 (2014)
25. Máyra Andressa, R.V., Piccioni, E.A.: Application of the finite element method in Dentistry. Rev. Sul-Brasil. Odontol. 369–377 (2013)
26. Viceconti, M., Bellingeri, L., Cristofolini, L., Toni, A.: A comparative study on different methods of automatic mesh generation of human femurs. Med. Eng. Phys. **20**(1), 1–10 (1998)
27. Austman, R.L., Milner, J.S., Holdsworth, D.W., Dunning, C.E.: The effect of the density–modulus relationship selected to apply material properties in a finite element model of long bone. J. Biomech. **41**(15), 3171–3176 (2008)
28. Kluess, D., Souffrant, R., Mittelmeier, W., Wree, A., Schmitz, K.-P., Bader, R.: A convenient approach for finite-element-analyses of orthopaedic implants in bone contact: modeling and experimental validation. Comput. Methods Prog. Biomed. **95**(1), 23–30 (2009)

29. Mughal, U.N., Khawaja, H.A., Moatamedi, M.: Finite element analysis of human femur bone. Int. J. Multiphys. **9**(2), 101–108 (2015)
30. Natali, A.N., Meroi, E.A.: A review of the biomechanical properties of bone as a material. J. Biomed. Eng. **11**(4), 266–276 (1989)
31. Hambli, R.: Connecting mechanics and bone cell activities in the bone remodeling process: an integrated finite element modeling. Front. Bioeng. Biotechnol **2** (2014)
32. Papini, M., Zdero, R., Schemitsch, E.H., Zalzal, P.: The biomechanics of human femurs in axial and torsional loading: comparison of finite element analysis, human cadaveric femurs, and synthetic femurs. J. Biomech. Eng. **129**(1), 12–19 (2006)
33. Lai, D.T.H., Begg, R.K., Palaniswami, M.: Computational Intelligence in gait research: a perspective on current applications and future challenges. IEEE Trans. Inf Technol. Biomed. **13**(5), 687–702 (2009)
34. Guiotto, A., Sawacha, Z., Guarneri, G., Avogaro, A., Cobelli, C.: 3D finite element model of the diabetic neuropathic foot: a gait analysis driven approach. J. Biomech. **47**(12), 3064–3071 (2014)
35. Christensen, R.: Analysis of Variance, Design and Regression. Chapman and Hall (1996)
36. Carpena, F.L., Tayo, L.L.: Finite element analysis of ACL reconstruction-compatible knee implant design with bone graft component. Computation **11**(8), 151 (2023)
37. Lewis, G.S., Mischler, D., Wee, H., Reid, J.S., Varga, P.: Finite element analysis of fracture fixation. Curr. Osteoporos. Rep. **19**, 403–416 (2021)
38. Lee, Y., Ogihara, N., Lee, T.: Assessment of finite element models for prediction of osteoporotic fracture. J. Mech. Behav. Biomed. Mater. **97**, 312–320 (2019)
39. Viceconti, M., Qasim, M., Bhattacharya, P., Imhauser, C.W., Monte, F.D.: Multiscale modelling of the musculoskeletal system: review of state of the art methods and future directions. Proc. Inst. Mech. Eng. **233**(3), 347–378 (2019)
40. Bansod, Y.D., Kebbach, M., Kluess, D., Bader, R., van Rienen, U.: Finite element analysis of bone remodelling with piezoelectric effects using an open-source framework. Biomech. Model. Mechanobiol. **20**(3), 1147–1166 (2021)
41. Wang, R., Wu, Z.: Recent advancement in finite element analysis of spinal interbody cages: a review. Front. Bioeng. Biotechnol. **11** (2023)
42. Yan, Z., et al.: Experimentally characterizing the spatially varying anisotropic mechanical property of cancellous bone via a Bayesian calibration method. J. Mech. Behav. Biomed. Mater. **138**, 105643 (2023)
43. Brekelmans, W.A., Poort, H.W., Slooff, T.J.: A new method to analyse the mechanical behaviour of skeletal parts. Acta Orthop. Scand. **43**(5), 301–317 (1972)
44. Cronin, D.S., Watson, B., Khor, F., Gierczycka, D., Malcolm, S.: Cortical bone continuum damage mechanics constitutive model with stress triaxiality criterion to predict fracture initiation and pattern. Front. Bioeng. Biotechnol. **10** (2022)
45. Giddings, V.L., King, M.T.: Challenges in patient-specific modeling: estimation of volumetric density from clinical imaging and the impact of the precision and resolution of 3D models in finite element simulations. J. Biomech. **73**, 131–137 (2018)
46. Zadpoor, A.A.: Patient-specific bone modeling and its potential for orthopedic surgery planning, bone functional adaptation, and bone regeneration. Mater. Today **18**(1), 66–73 (2015)
47. González-Carbonell, R.A., Ortiz-Prado, A., Cisneros-Hidalgo, Y.A., Alpizar-Aguirre, A.: Bone remodeling simulation of subject-specific model of tibia under torque. In: Braidot, A., Hadad, A. (eds.) VI Latin American Congress on Biomedical Engineering CLAIB 2014, Paraná, Argentina 29, 30 & 31 October 2014. IP, vol. 49, pp. 305–308. Springer, Cham (2015). https://doi.org/10.1007/978-3-319-13117-7_79

48. Hambli, R., Katerchi, H., Benhamou, C.-L.: Multiscale methodology for bone remodelling simulation using coupled finite element and neural network computation. Biomech. Model. Mechanobiol. **10**(1), 133–145 (2010)
49. Guha, I., Zhang, X., Rajapakse, C.S., Chang, G., Saha, P.K.: Finite element analysis of trabecular bone microstructure using CT imaging and continuum mechanical modeling. Med. Phys. **49**(6), 3886–3899 (2022)

Thermodynamic Study of Novel Fe-Mn-Si Alloys

Suresh Chand[1,2](✉), Khushboo Rakha[2], and Ravi Mohan Prasad[2]

[1] Punjab Engineering College, Chandigarh, India
sureshchand@pec.edu.in
[2] Indian Institute of Technology Ropar, Rupnagar, India

Abstract. With the emergence of advanced high-strength alloys for the automotive industry, the utilization of advanced high-strength alloys is considered for the replacement of existing alloy steels used in the automotive industries. To develop advanced high-strength alloys, there is a need to understand the thermodynamic behavior of the alloy systems. In this paper, the thermodynamic behavior (entropy, free energy, and enthalpy) of Fe-xMn-1Si (x = 3–5 wt.%) alloys are studied.

Keywords: Thermodynamics · phases · steels · Medium manganese · thermo-Calc

1 Introduction

Medium manganese steels, with manganese percentages in the range of 4–12 weight percent, are being considered as potential alloys for the third generation of advanced high-strength steels (AHSS). This is due to their ability to produce increased strength and ductility while maintaining cost-effectiveness in terms of materials. The approach is employed in the development of medium-manganese (medium-Mn) steels incorporating aluminum and/or nickel. Utilizing computational alloy thermodynamics, an extensive range of compositions was methodically explored to chart the proportion and stability of retained austenite at intercritical annealing temperature [1]. The process of ferrite formation during the standard phase transformation is characterized by a slow transformation. However, the concept of composition design, which involves incorporating elements that stabilize austenite, serves as a valuable approach for achieving high strength through the formation of a low-temperature phase (martensite). This design strategy enables the production of medium-manganese high-strength steels on a hot continuous rolling line, allowing rapid austenite formation and customization of retained proportion and stability [2]. Demonstrations have revealed the potential for achieving remarkable combinations of strength and elongation through the activation of twinning-induced plasticity and transformation-induced plasticity effects in a steel composition of (0.15–0.30)C-60Mn-1.50Si-30Al (wt%). The key to achieving favourable mechanical properties lies in adjusting the room temperature stacking fault energy of austenite through careful selection of annealing temperatures. This adjustment was found to be crucial in obtaining desirable mechanical characteristics. The tensile behaviour of medium-manganese steels containing δ-ferrite and constituents of austenite and α-ferrite was explored through a

A. Tiwari et al. (Eds.): iConFIRM 2023, SPPHY 308, pp. 82–91, 2024.
https://doi.org/10.1007/978-981-97-4557-9_9

combination of thermodynamic modelling and comprehensive characterization, along with mechanical testing [3]. The study used computational modeling to analyze the transformation of martensite-to-austenite and alloying element distribution in a steel composition with 0.18% carbon, 11% manganese, and 3.8 wt.% aluminum [4]. The study investigated austenite growth from ferrite and martensite between the A1 and A3 regions of medium manganese steels using computational kinetics, H. Farahani et al. [5]. F. Huyan et al.'s computational study analyzed austenite growth from martensite, examining carbide precipitation, dissolution, and early stages in medium manganese steels [6]. Researchers are working on a comprehensive alloy and process design framework for medium manganese steels using advanced techniques like genetic algorithms, plural-objective optimization, and machine learning in conjunction with CALPHAD and FEM modeling [7]. The multi-objective optimization scheme, computational thermodynamics, and grid search algorithm were used to develop Ni-containing medium Mn steels, improving formability and reducing manufacturing costs, using a systematic Integrated Computational Materials Engineering strategy [8]. A new thermodynamic model predicts austenite's stacking fault energy, considering elements like C, Mn, Al, Si, and Cr, and particle size [9]. The present work emphasizes the thermodynamic study in terms of thermodynamic parameters.

2 Materials and Thermodynamic Models

The thermodynamic study of (Fe-(3-5) Mn-1Si) medium manganese steel with silicon is carried out in thermo-Calc software. The compositions of the alloys studied are presented in Table 1. The study of enthalpy, entropy, Gibbs free energy, and specific phase diagrams is performed.

The Gibbs free energies of pure elements concerning temperature are represented by the following equation:

$$\begin{aligned} {}^{0}\mathrm{Gi}(\mathrm{T_1}) &= \mathrm{a} + \mathrm{bT_1} + \mathrm{cT_1} \times \ln(\mathrm{T_1}) + \mathrm{dT_1^2} + \mathrm{eT_1^{-1}} \\ &\quad + \mathrm{fT_1^3} + \mathrm{iT_1^4} + \mathrm{jT_1^7} + \mathrm{kT_1^{-9}} \end{aligned}$$

The ${}^{0}\mathrm{Gi}(\mathrm{T_1})$ data refers to the constant Gibbs free energy value of the standard element reference (SER) at 298.15 K and 1 bar adopted from Scientific Group Thermodata Europe.

The Gibbs energy of the liquid, FCC (A1), BCC (A2), CBCC (A12), CUB (A13), and HCP (A3) phases can be expressed with the substitutional solid solution model as:

$$G_m^{\Phi} = \sum_{i=Fe,Mn,Si} x_i {}^{o}G_i^{\Phi} + {}^{ex}G_m^{\Phi} + RT \times \sum_{i=Fe,Mn,Si} x_i \ln(x_i)$$

where x_i is the elemental mole fraction. T is the absolute temperature, while R is the gas constant. ${}^{o}G_i^{\Phi}$ is the molar Gibbs energy of the pure state in the standard state. ${}^{ex}G_m^{\Phi}$ is molar excess free energy, which is given by

$$\begin{aligned} {}^{ex}G_m^{\Phi} &= x_{Fe}.x_{Mn}L_{Fe,Mn}^{\Phi} + x_{Mn}.x_{Si}L_{Mn,Si}^{\Phi} + x_{Si}.x_{Fe}L_{Si,Fe}^{\Phi} \\ &\quad + x_{Fe}.x_{Mn}x_{Si}L_{Fe,Mn,Si}^{\Phi} \end{aligned}$$

Redlich Kister polynomials created binary interaction parameters, $L^{\Phi}_{i,j}$ expressed in the above equation

$$L^{\Phi}_{i,j} = \sum_{n} (x_i - x_j) \cdot {}^{n}L^{\Phi}_{i,j}$$

where ${}^{n}L^{\Phi}_{i,j} = a + bT$

The interaction parameter for 3 element system $L^{\Phi}_{Fe,Mn,Si}$ is expressed as under

$$L^{\Phi}_{Fe,Mn,Si} = x_{Fe} \cdot {}^{0}L^{\Phi}_{Fe,Mn,Si} + x_{Mn} \cdot {}^{1}L^{\Phi}_{Fe,Mn,Si} + x_{Si} \cdot {}^{2}L^{\Phi}_{Fe,Mn,Si}$$

The Gibbs energy equation of intermetallics is expressed as

$$G_m^{(Fe,Mn)_1Si_1} = y_{Fe} \cdot {}^{0}G_{Fe:Si}^{(Fe,Mn)_1Si_1} + y_{Mn} \cdot {}^{0}G_{Mn:Si}^{(Fe,Mn)_1Si_1}$$
$$+RT(y_{Fe}\ln_{Fe} + y_{Mn}\ln_{Mn}) + y_{Fe} \cdot y_{Mn} {}^{0}G_{Fe,Mn:Si}^{(Fe,Mn)_1Si_1}$$

where y_{Fe} and y_{Mn} are mole fractions of Fe and Mn in the first sublattice, respectively. ${}^{0}G_{Fe:Si}^{(Fe,Mn)_1Si_1}$ and are the Gibbs energies of (FeSi) in the Fe-Si, and MnSi in Mn-Si alloys. ${}^{0}L_{Fe,Mn:Si}^{(Fe,Mn)_1Si_1}$ is the interaction for Fe and Mn [9–11].

Table 1. Compositions and transformation temperatures

Alloy/Elements	Mn (wt%)	Si (wt%)	Fe (wt%)	A1 (°C)	A3 (°C)
Alloy A	3	1	96	695	843
Alloy B	4	1	95	678	821
Alloy C	5	1	94	661	793

3 Results and Discussion

Figure 1 shows the phase diagrams developed using thermo-Calc software. Figure 1(a) shows the phase fields with variations in manganese contents. While Fig. 1(b) represents a phase field with an increase in silicon contents in the alloy. The Gibbs free energy diagrams at 1000 K temperature for Fe-Mn and Fe-Si are shown in Fig. 2. The Gibbs free energy curve for the austenite phase (FCC_A1) is most negating in Fig. 2, it indicates that the austenite phase is the most stable while other phases present at that temperature have lower relative stability. The relative stability of α-manganese (BCC_A1), β-manganese (cubic), and other phases is shown in Fig. 3. The star marks represent Gibbs free energy for possible intermetallics.

The activity of iron decreases in the presence of silicon, while the activity of silicon is much less, as shown in Fig. 4(a). The activity of iron decreases with increasing manganese contents, while the activity of manganese increases with increasing Mn contents

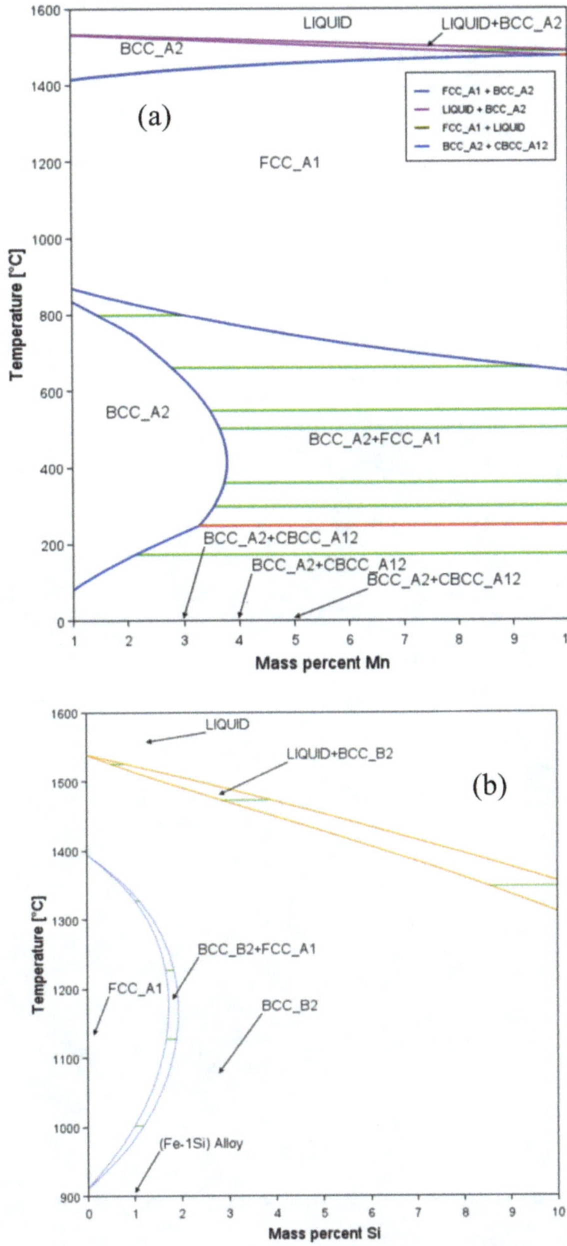

Fig. 1. Alloy phase diagrams, (a) Fe-Mn phase diagrams, (b) Fe-Si phase diagrams.

in the alloys. The volume fractions of cementite and intermetallics are less than 5 percent. The volume fractions of austenite and ferrite are in equal proportion at temperatures slightly lower than 1000 K, as shown in Fig. 5. The slow nature of ferrite formation in the standard phase transition highlights the importance of employing austenite stabilizer

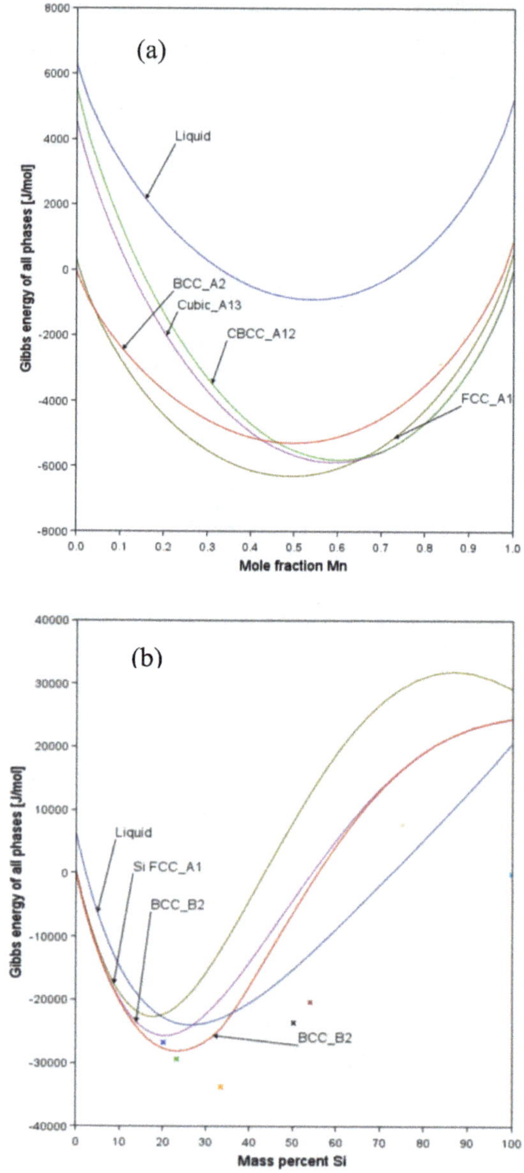

Fig. 2. Gibbs free energy of phases at 1000 K, (a) Fe-Mn system, (b) Fe-Si system.

elements in the compositional design. This strategy offers valuable direction for producing medium manganese high-strength alloys with a focus on achieving high strength in the low-temperature phase (martensite). In contrast, the reverse phase transformation involving austenite formation occurs relatively swiftly. The initial martensite and reversion parameters can be adjusted to customise the vol.% and stability of retained austenite. This method is well-suited for developing medium-Mn alloys with stringent

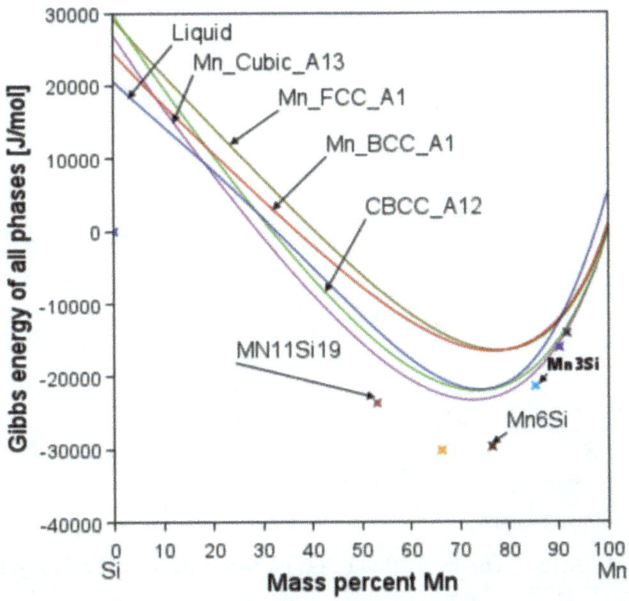

Fig. 3. Gibbs free energy of phases at 1000 K, Si-Mn system.

requirements for both strength and ductility [2]. The effect of manganese on enthalpy and entropy at 1000 K temperature is shown in Fig. 6. The enthalpy decreases by increasing Mn contents up to 4 wt.% while it starts decreasing by increasing Mn. The entropy increases by increasing the amount of Mn. The degree of randomness will increase by increasing Mn in the alloy. The enthalpy increases with an increase in temperature. The increase in enthalpy is found to be non-linear in two locations on the plot shown in Fig. 7. The enthalpy of alloys increases with increasing temperature (900–1100 °C), which is α-ferrite (BCC-A2) to γ-austenite (FCC_A1) transformations. A non-linear curve after a 1700 K sudden rise in enthalpy indicates the melting of the alloy.

The entropy of alloys increases with increasing temperature in two non-linear regions of the plots, as shown in Fig. 8.

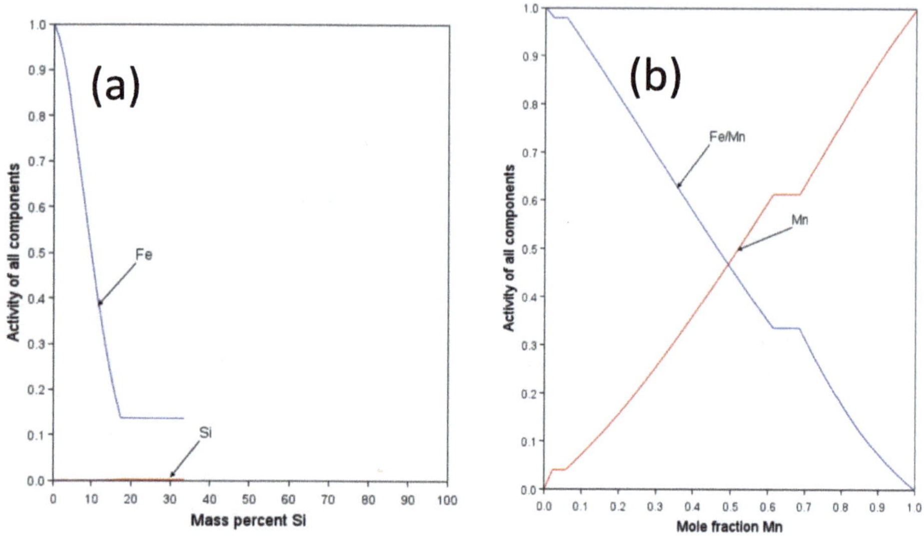

Fig. 4. Activity curves at 1000 K, (a) Fe-Mn system, (b) Fe-Si system.

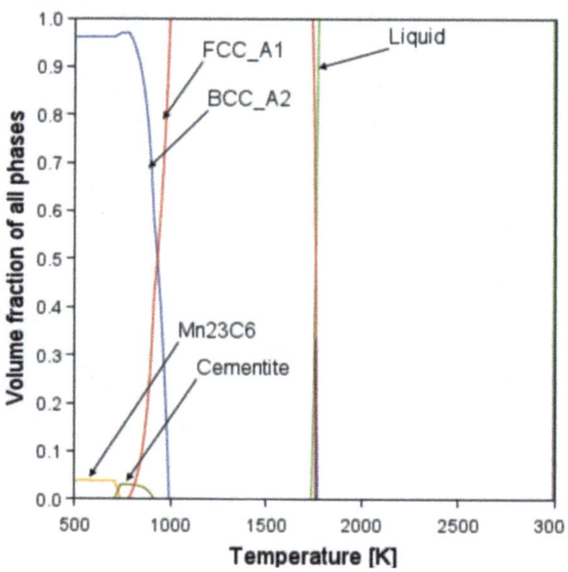

Fig. 5. Volume fraction of phases Fe-4Mn-1Si system.

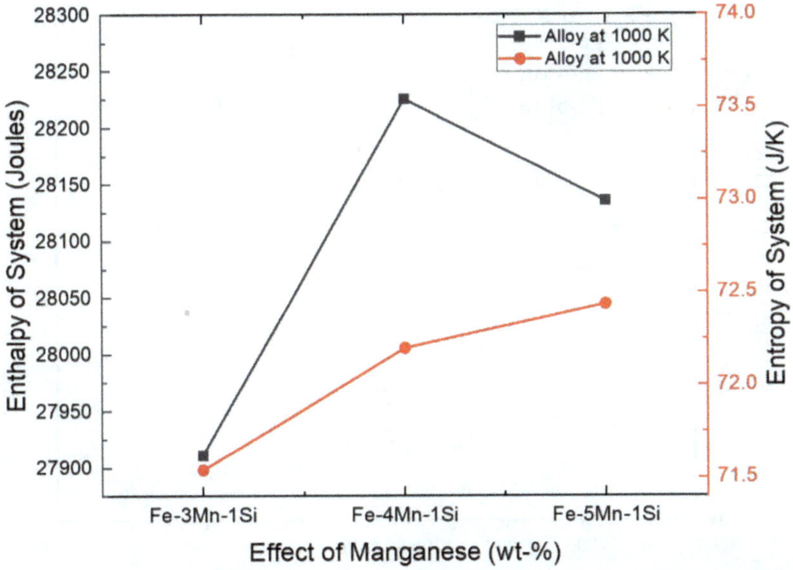

Fig. 6. Enthalpy (dark curve) and Entropy (red curve) of medium manganese alloys. (Color figure online)

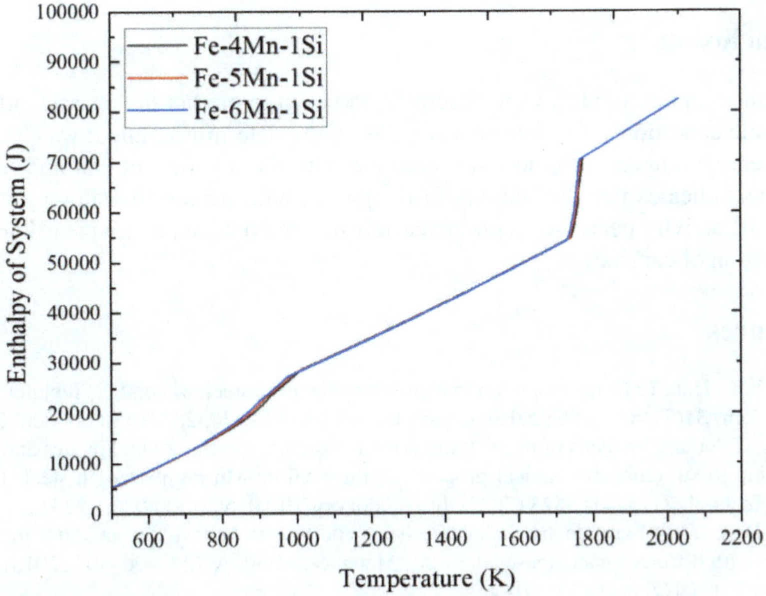

Fig. 7. Enthalpy of different alloy systems: Fe-3Mn-1Si system-dark curve, Fe 4Mn-1Si, Fe-5Mn-1Si system-blue curve. (Color figure online)

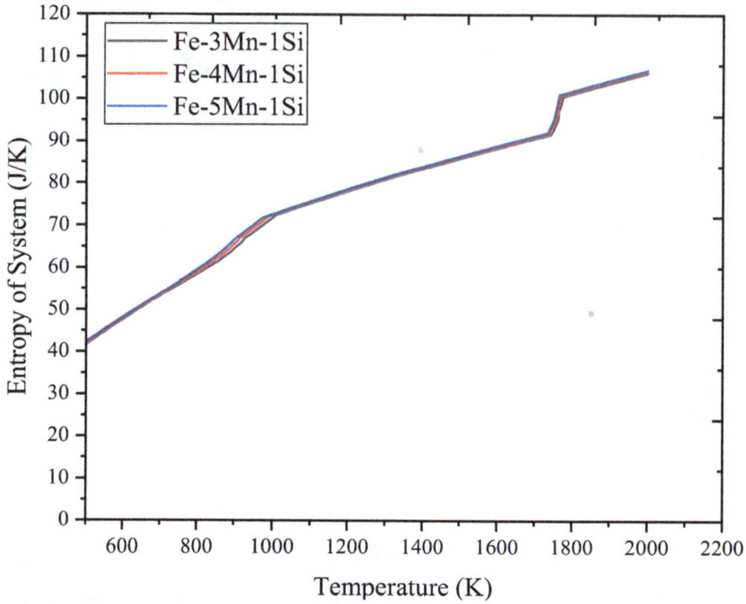

Fig. 8. Entropy of different alloy systems: Fe-3Mn-1Si system-dark curve, Fe4Mn-1Si system-red curve, Fe-5Mn-1Si system-blue curve (Color figure online)

4 Conclusion

The thermodynamic models were described and used to predict the phases under thermodynamic conditions. The free energy curves were determined, and it was found that the free energy curves become more negative with the addition of Mn and Si in the alloy. This indicates that the stability of the phases will increase by alloying with Mn and Si. The activity decreases with increasing silicon contents, consequently lowering the formation of carbides.

References

1. Lee, Y.K., Han, J.: Current opinion in medium manganese steel. Mater. Sci. Technol. (United Kingdom) **31**(7), 843–856 (2015). https://doi.org/10.1179/1743284714Y.0000000722
2. Dong, Y., et al.: Analysis of phase transformation thermodynamics and kinetics and its relationship to structure-mechanical properties in a medium-Mn high strength steel. J. Mater. Res. Technol. **27**, 5411–5423 (2023). https://doi.org/10.1016/j.jmrt.2023.10.231
3. Sun, B., et al.: Microstructural characteristics and tensile behavior of medium manganese steels with different manganese additions. Mater. Sci. Eng. A **729**, 496–507 (2018). https://doi.org/10.1016/j.msea.2018.04.115
4. Kamoutsi, H., Gioti, E., Haidemenopoulos, G.N., Cai, Z., Ding, H.: Kinetics of solute partitioning during intercritical annealing of a medium-mn steel. Metall. Mater. Trans. A Phys. Metall. Mater. Sci. **46**(11), 4841–4846 (2015). https://doi.org/10.1007/s11661-015-3118-7
5. Farahani, H., Xu, W., van der Zwaag, S.: Prediction and validation of the austenite phase fraction upon intercritical annealing of medium Mn steels. Metall. Mater. Trans. A Phys. Metall. Mater. Sci. **46**(11), 4978–4985 (2015). https://doi.org/10.1007/s11661-015-3081-3

6. Huyan, F., Yan, J.Y., Höglund, L., Ågren, J., Borgenstam, A.: Simulation of the growth of austenite from as-quenched martensite in medium Mn steels. Metall. Mater. Trans. A Phys. Metall. Mater. Sci. **49**(4), 1053–1060 (2018). https://doi.org/10.1007/s11661-018-4497-3
7. Khan, D., Shukla, R., Gautham, B.P.: In silico design of materials and processes: an application of ICME to carburizing steels. Trans. Indian Inst. Met. **72**(8), 2179–2185 (2019). https://doi.org/10.1007/s12666-018-1534-2
8. Aristeidakis, J.S., Haidemenopoulos, G.N.: Alloy design based on computational thermodynamics and multi-objective optimization: the case of medium-Mn steels. Metall. Mater. Trans. A Phys. Metall. Mater. Sci. **48**(5), 2584–2602 (2017). https://doi.org/10.1007/s11661-017-4010-4
9. Aristeidakis, J.S., Haidemenopoulos, G.N.: Composition and processing design of medium-Mn steels based on CALPHAD, SFE modeling, and genetic optimization. Acta Materialia (2020). https://doi.org/10.1016/j.actamat.2020.03.052
10. Dinsdale, A.T.: SGTE data for elements. Calphad **15**(4), 317–425 (1991)
11. Forsberg, A., Ågren, J.: Thermodynamic evaluation of the Fe-Mn-SI system and the γ/ε martensitic transformation. J. Phase Equilibria **14**(3), 354–363 (1993)

Hydride Induced Degradation of Mechanical Properties of Zr-2.5Nb Alloy

R. N. Singh(✉)

Outstanding Scientist and Head, Mechanical Metallurgy Division, Modular Laboratories, Bhabha Atomic Research Centre, Trombay, Mumbai, India
rnsingh@barc.gov.in

Abstract. Tubes made of Zr-2.5Nb alloy are used in pressurized heavy water reactors as pressure boundaries for hot coolant and are called pressure tubes, which are subjected to slow aqueous corrosion resulting in nascent hydrogen generation. Though most of the nascent hydrogen evolved recombines to form molecular hydrogen and is unable to enter into the pressure tubes, a part of the hydrogen is picked up by the pressure tube material. Once the solid solubility is exceeded brittle hydrides of plate-shaped morphology form that make the host matrix brittle. Two forms of hydride embrittlement have been recognized for Zr-alloys. These are gross and localized hydride embrittlement. For the manifestation of gross hydride embrittlement, a minimum volume fraction of hydride is required. The degree of gross hydride embrittlement is significantly affected by the hydride plate orientation. The localized form of hydride embrittlement occurs for hydrogen concentration below the solid solubility limit and is caused by hydrogen migration to a localized spot. Once the local solid solubility is exceeded hydride precipitation takes place. Various aspects of hydride embrittlement of Zr-alloys and length scales involved are discussed.

Keywords: Zr-2.5Nb alloys · Hydride Embrittlement · Stress Free Transformation Strain · Delayed Hydride Cracking · Leak Before Break · Critical Crack Length

1 Introduction

Zr-2.5Nb alloy is used as a pressure boundary for hot coolant in pressurized heavy water reactors [1–4]. The reaction of hot coolant heavy water with zirconium produces nascent hydrogen. A small fraction of hydrogen thus produced enters the pressure tube material. Hydrogen is soluble in zirconium metal and the terminal solid solubility defined as the maximum amount of hydrogen that can be retained in solid solution, increases exponentially with an increase in temperature [5, 6]. Hydrogen present in the zirconium metal more than the solid solubility precipitates in the form of hydride. Four types of hydrides can form in dilute zirconium alloys [1]. However, for conditions prevailing in the reactor δ–hydride of plate shape forms in the Zr-2.5Nb pressure tube material [1, 7]. The δ–hydride is brittle and makes the host matrix brittle and this phenomenon is called hydride embrittlement.

A. Tiwari et al. (Eds.): iConFIRM 2023, SPPHY 308, pp. 92–100, 2024.
https://doi.org/10.1007/978-981-97-4557-9_10

The degree of hydride embrittlement is affected to different extents by hydrogen concentration, temperature, direction of approach to temperature, nature of hydride, hydride orientation, size of hydride precipitate and loading rate [1, 7]. Two forms of hydride embrittlement have been recognized in the literature. These are gross and localized forms of embrittlement. Gross hydride embrittlement is caused by a certain minimum volume fraction of hydride phase, results in a reduction in tensile ductility, impact and fracture toughness and is significantly affected by hydride orientation [1, 7]. The localized form is caused by hydrogen migration under chemical potential gradient and manifests under concentration, hydrostatic stress or temperature gradients either individually or in combination [1, 7].

Hydride embrittlement in its entirety is a complex problem involving thermodynamics and kinetics, phase transformation, deformation, diffusion, solid mechanics and fracture mechanics, and in many instances the processes are coupled and manifest at different length scales [1]. Both experimental and computational approaches are used to understand various aspects of hydride embrittlement [1]. The mechanical properties of the pressure tube material are used for its safety assessment and the extent of degradation in mechanical properties due to hydride embrittlement is used for life estimation and extension [1]. In this article select facets of hydride embrittlement of Zr-alloys manifested at different length scales are discussed.

2 Atomic Scale–Stress Free Transformation Strain

Goldak et al. [8] reported the lattice parameters of hcp zirconium phase in 1966 as a function of temperature. Bradbrook et al. [9] reported the lattice parameter of δ–hydride in 1972. Carpenter [10] in 1973 assumed ellipsoidal and spherical shape of zirconium atom in hcp and fcc phases, respectively, and used the atomic radius of zirconium in hcp α–Zr phase and δ–hydride to determine the stress-free transformation of hydrides in zirconium alloys at ambient temperature. Yamanaka et al. [11] reported the lattice parameter of hydride from ambient to 500 °C in 2005 and using the values of lattice parameter and following Carpenter's methodology Singh et al. [12] calculated the stress-free transformation strain of δ–hydride between 25 to 500 °C in 2007. Figure 1 shows the temperature dependence of lattice parameters of α–Zr, δ–hydride and stress-free transformation strain components e_{11} and e_{22} [12].

The δ–hydride formation is accompanied by an increase in volume of ~17% at 25 °C and is observed to increase with an increase in temperature [12]. Such a large change in volume causes local plastic deformation of the matrix. The minimization of the total strain energy of the Zr-H-hydride system computed by a finite element method has been used by Singh et. al. [13] and Sharma et al. [14] to predict the shape and habit plane of hydride. Figure 2 shows the variation of strain energy of δ–hydride with aspect ratio defined as the ratio of the height and breadth of the hydride plate. The legends representing various combinations of matrix and hydride properties are described in Table 1. As can be seen from this figure the accommodation strain energy is observed to decrease with a decrease in aspect ratio. However, as the aspect ratio decreases the interfacial energy increases and optimum shape of hydride is decided by both accommodation energy and interfacial energy.

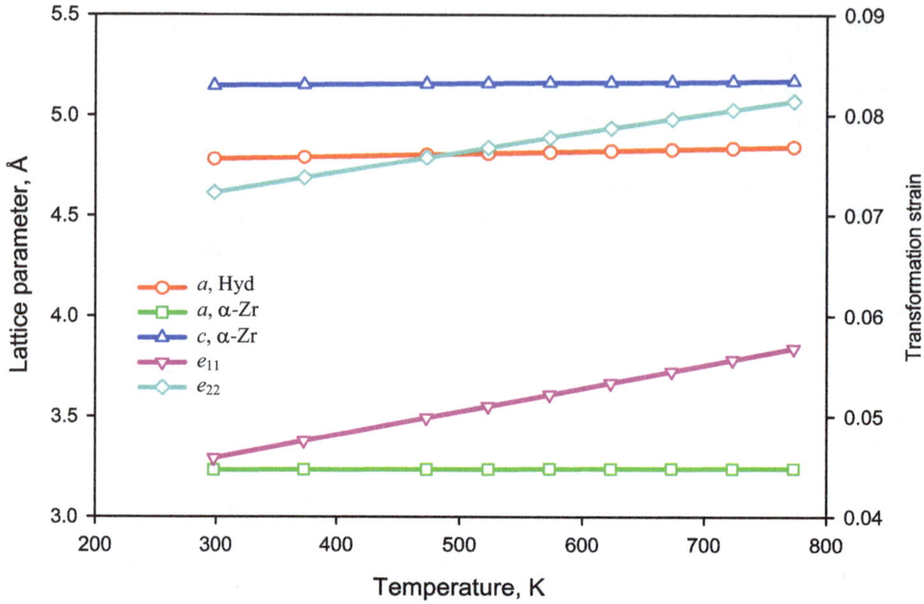

Fig. 1. Plot of lattice parameters of α–Zr and δ–hydride, and stress-free transformation strain components e_{11} and e_{22} of δ–hydride between 298 to 773 K [12].

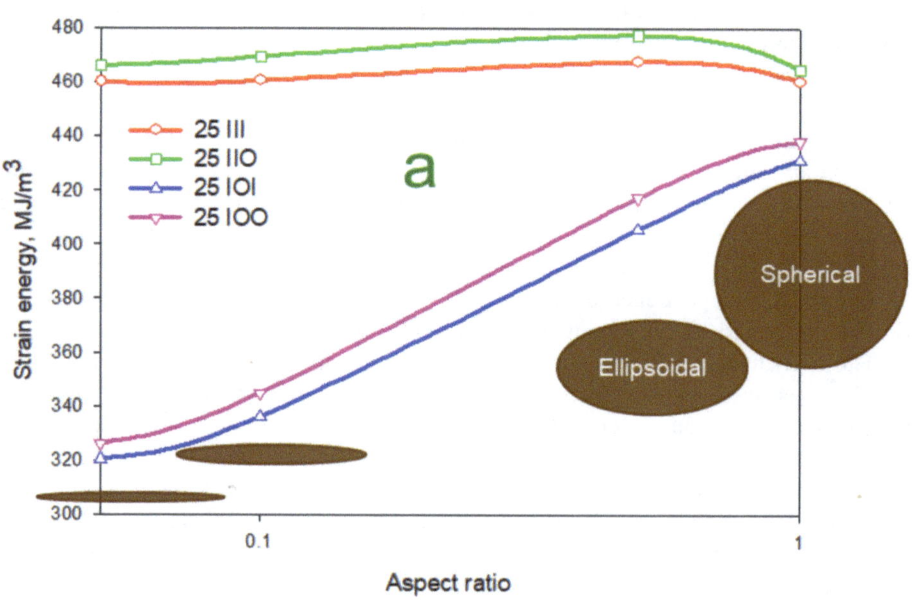

Fig. 2. Plot of accommodation strain energy versus the aspect ratio of hydride for fully elastic matrix and hydride.

Table 1. Various combinations of material properties used for computation of accommodation strain energy of hydride.

Case	Hydride modulus	Transformation strain	Matrix modulus	Strain energy
III	Isotropic	Isotropic	Isotropic	No reduction
IIO	Isotropic	Isotropic	Orthotropic	No significant reduction
IOI	Isotropic	Orthotropic	Isotropic	Significant reduction
IOO	Isotropic	Orthotropic	Orthotropic	Significant reduction

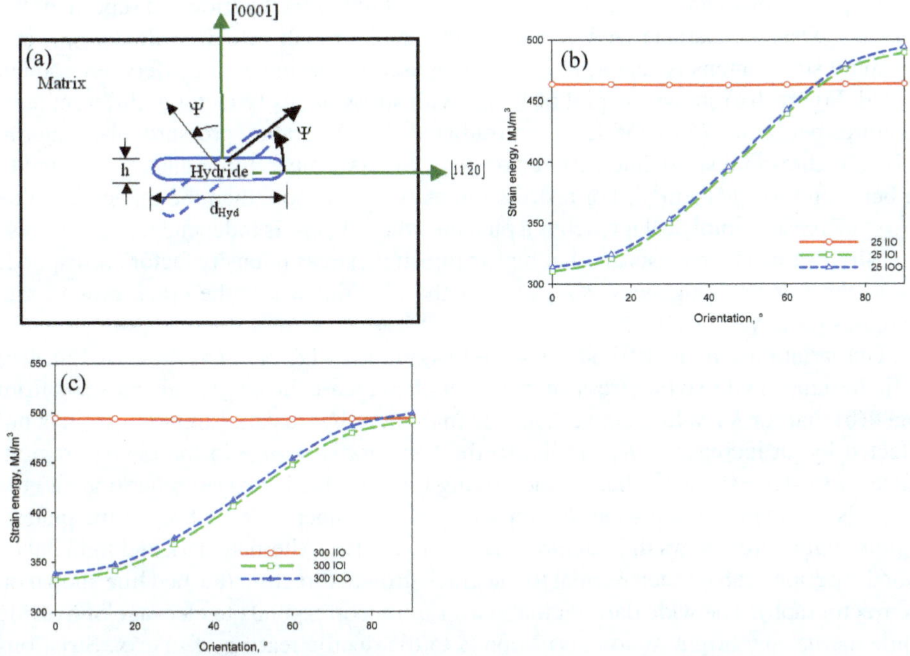

Fig. 3. (a) Sketch showing hydride plate inside hcp zirconium crystal. Plot of accommodation strain energy with orientation of hydride for fully elastic matrix and hydride obtained using the properties at (b) 25 °C and (c) 300 °C.

The habit plane of the hydride can also be found by minimization of the accommodation strain energy. Figure 3(a) shows α–hcp zirconium crystal with basal pole oriented vertically. A δ–hydride plate is also shown in this figure. The plot of accommodation strain energy with hydride orientation obtained using elastic material properties of matrix and hydride at 25 °C and 300 °C is shown in Fig. 3(b) and (c), respectively. Zero degree orientation corresponds to a hydride plate oriented along the basal plane of the hcp zirconium. As is evident from this figure the accommodation strain energy of the hydride shows a minimum when the hydride plate is oriented parallel to the basal plane. This suggests that the basal plane of hcp zirconium is the microscopic habit plane of hydride at both 25 and 300 °C. Similar results were obtained using elasto-plastic material properties of matrix and δ–hydride.

3 Microscopic Scale – Delayed Hydride Cracking

Delayed hydride cracking (DHC) is a localized form of hydride embrittlement [1, 7, 15–17]. The condition for DHC to occur is a hydrogen concentration more than a threshold concentration, a loaded flaw with a stress intensity factor more than a critical value and temperature in between a lower and upper bound value. It is caused by hydrogen migration up the hydrostatic stress gradient resulting in hydride precipitation near the crack tip, its growth and final crack growth by cracking of the hydride and rupture of the ligament. DHC is characterized by an average crack growth velocity called V_{DHC} and threshold stress intensity factor, K_{IH}, which is used by designers for safety assessment [1, 16]. Figure 4(a) shows the plot of V_{DHC} with stress intensity factor at different temperatures between 476 to 556 K for unirradiated Zr-2.5Nb alloy pressure tube material [17]. The threshold stress intensity factor, K_{IH}, in this temperature range, was found to be between 9–11 MPa.m$^{1/2}$. For a stress intensity factor more than the threshold value V_{DHC} increases rapidly till it reaches a plateau, where V_{DHC} is independent of the stress intensity factor. There exists another higher threshold stress intensity factor corresponding to the fracture toughness, K_{IC}, [18] of the material where the crack growth rate increases rapidly.

The variation of hydrostatic stress with stress intensity factor, K_I, is shown in Fig. 4(b) [17]. Initially, hydrostatic stress increases with increase in K_I. It can be seen from Fig. 4(b) that for K_I values above K_{IH}, the maximum hydrostatic stress ($\sigma_{h,\max}$), is not affected by an increase in K_I similar to the behavior of V_{DHC} in the plateau region. Since hydrostatic stress gradient is the driving force for the hydrogen migration and the $\sigma_{h,\max}$ is constant in the plateau region V_{DHC} is also independent of K_I in the plateau region. Figure 4(c) shows the fracture surface [1, 18, 19]. Alternate dark and light bands called striations can be seen normal to the crack growth direction (dashed line shown on the fractograph). The wide dark region in a striation corresponds to cleavage of hydride while the narrow bright region corresponds to the ductile tearing of matrix. Striations are evidence of a discontinuous nature of crack growth during delayed hydride cracking

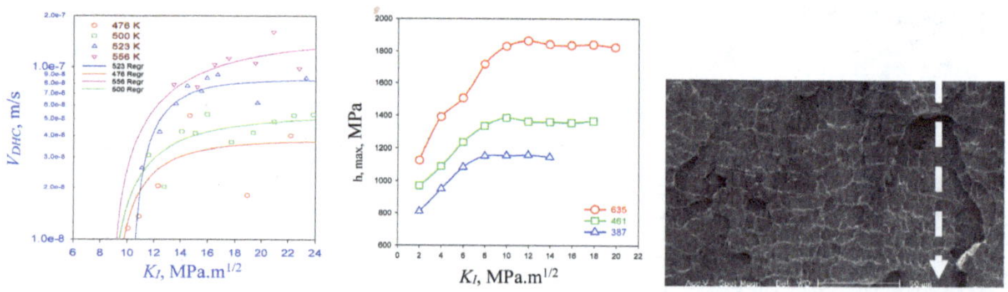

Fig. 4. Plot of (a) V_{DHC} and (b) hydrostatic stress with stress intensity factor (K_I) for Zr-2.5Nb alloy [18]. The hydrostatic stress was computed assuming different yield strength (MPa) of the material. (c) Fracture surface having striations as evidence of a discontinuous nature of crack growth during DHC.

and are related to plastic zone size. With an increase in strength, striation spacing is observed to decrease [19].

4 Mesoscopic Scale – Fracture

Fracture toughness tests are performed as per ASTM-E1820 [20] using 17 mm width curved compact toughness specimens [18]. The crack length is determined either by unloading compliance [21] or DCPD techniques [18]. Fracture toughness parameters like J_Q, J_{max}, and *dJ/da* are determined. The variation in fracture toughness parameter J_{max}, which corresponds to a value of *J* at maximum load, with test temperature is shown in Fig. 5(a) for different hydrogen concentrations for double melted Zr-2.5Nb pressure tube material [18]. It is evident that an increase in hydrogen concentration from 20 and 100 wppm is not affecting the fracture toughness of the double-melted material [18]. The variation in J_{max} with deuterium concentration expressed in hydrogen equivalent [22] at different temperatures is shown in Fig. 5(b) for quadruple-melted Zr-2.5Nb pressure tube material. At 25 °C, it can be that as the deuterium concentration increases from 5 to 30 wppm there is a drastic reduction in fracture toughness and with further increase in deuterium concentration no change in J_{max} is observed. With an increase in temperature the extent of reduction J_{max} reduces.

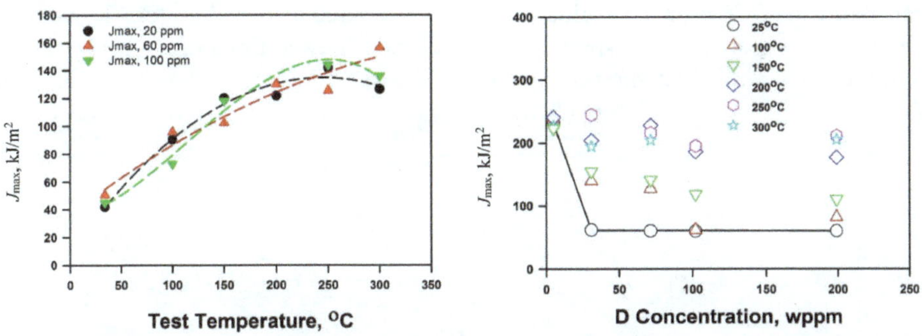

Fig. 5. Variation J_{max} with (a) test temperature for double-melted [18] and (b) deuterium concentration for quadruple-melted Zr-2.5Nb alloy pressure tube material [22].

5 Macroscopic Scale – Component Failure

The pressure tube design is based on the concept of leak before break (LBB) [1]. Level 1 of the LBB criteria requires that there should not be any flaw in the component. Level 2 of LBB criteria stipulates that the any existing flaw should not grow during service. Level 3 requires that if any existing flaw grows during service it should cause leak rather than break. For adherence to LBB criteria the value of critical crack length (*CCL*), leak size crack a_{Leak}, V_{DHC} and operator response time (*t*) is required [1]. The operator response time should be lower than the time required for a leaking crack to grow to *CCL*.

The CCL is determined by using fracture toughness parameter J_{IC} or J_Q or J_{max} and operating parameters using Eq. 1.

$$J_Q = \frac{8a\sigma_{flow}^2}{\pi E} \ln\left[\sec\left(\frac{\pi M S_H}{2\sigma_{flow}}\right)\right] \tag{1}$$

where, $a = 1/2\,\mathrm{CCL}$ and $S_H = \frac{PD}{2B}$, σ_{flow} is the flow stress of the material, E is the Young's modulus, S_H is the hoop stress and M is Folia's factor. In Eq. 1 the term on left-hand side represents fracture toughness parameter and the term of right-hand side represents crack driving force under operating condition. The CCL is affected by hydrogen content, hydride orientation, microstructure and texture and hoop stress, flow stress, Young's modulus and component geometry.

The plot of CCL with test temperature for unirradiated Zr-2.5Nb pressure tube material in the as-received and hydrogen charged condition is shown in Fig. 6 [23]. The hydrogen was charged gaseously in the samples using a modified Sievert's apparatus [24, 25] and the hydrogen concentration analyzed by inert gas fusion technique was found to be 100 wppm [23]. As can be seen from this figure the CCL for the as-received (AR) material is not affected by temperature. The CCL for the materials in the as-charged condition contained circumferential hydride and at 25 °C the CCL is about 30% lower than the values for the AR material. With increase in temperature the CCL for the material containing circumferential hydride was observed to increase and at reactor operating temperature the CCL is only marginally lower as compared to the values for AR material. The material containing 100 wppm of radial hydride showed three regions, viz., lower shelf, transition and upper shelf regions [23]. The transition from lower to upper shelf

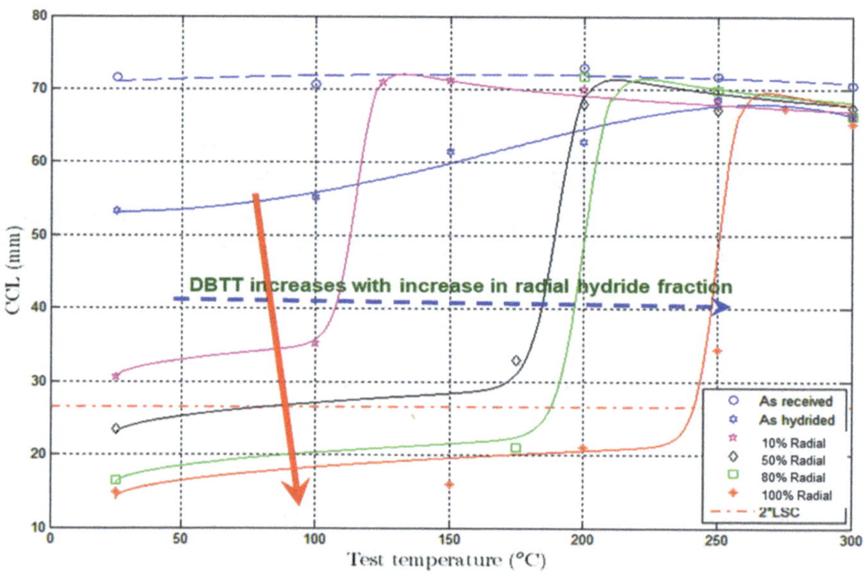

Fig. 6. Plot of critical crack length (CCL) with temperature for unirradiated Zr-2.5Nb alloy in the as received, as hydride and for various fraction of radial hydride [23].

is very sharp and is observed to increase with increase in radial hydride fraction. For the material containing fully radial hydride the lower shelf temperature extends close to 250 °C, which is comparable to the coolant reactor inlet temperature [23].

The *CCL* and V_{DHC} is required for validating LBB Criteria using Eq. 2. The operator response time (t) is about 12 h [1]. The *CCL* and V_{DHC} are obtained under operating condition as described earlier. The denominator 2 on right hand side accounts for bi-directional growth of crack in axial direction due to DHC.

$$t < \frac{CCL - a_{Leak}}{2V_{\mathrm{DHC}}} \tag{2}$$

If a leaking crack size is less than *CCL*, LBB will occur while for a leaking crack size greater than *CCL*, break before leak will occur.

6 Summary

Multi-scale Structural mechanics starting from atomic level to component level was illustrated through various aspects of hydrogen related problems of Zr-alloys. Using this approach not only the underlying degradation mechanism can be understood but valuable input for designer and safety engineers can be obtained.

Acknowledgements. Constant encouragement and invaluable support provided by Dr. A. K. Mohanty, Secretary, Department of Atomic Energy & Chairman, Atomic Energy Commission, Mr. Vivek Bhasin, Director, BARC and Dr. R. Tewari, Director, Materials Group, BARC is acknowledged. Author is grateful to Dr. S. Banerjee and Dr. R. K. Sinha former secretaries, Department of Atomic Energy, Prof. B. P. Kashyap, Ph.D. Supervisor, Prof. Per Ståhle von Schwerin, MC-IIF host and Mr. S. A. Bhardwaj former Chairman AERB for their guidance. The support from Dr. A. K. Suri, Dr. J. K. Chakravartty, Dr. G. K. Dey, Dr. Madangopal K. and Dr. V. Kain former Directors, Materials Group is acknowledged. Thanks are due to all our colleagues from BARC, NPCIL and NFC whose contributions have been cited in this article.

References

1. Singh, R.N., Bind, A.K., Sunil, S., Sarkar, A., Neogy, S., Murty, T.N.: On the evolution of a zirconium alloy for use as pressure tubes in Indian pressurized heavy water reactors. In: Srivatsan, T.S., Karloopia, J.,Gupta, M., Eds., Chapter 5 in the Book titled 'Advanced Materials for Emerging Applications. Bentham Science Publishers (2023)
2. Singh, R.N., Chaterjee, S., Banerjee, S.: PHWR coolant channel safety issues – a metallurgical perspective. In: First National Conference on 'Nuclear Reactor Technology – NRT-1' Nuclear Reactor Safety, Held on November 25-27th, 2002, BARC, Mumbai, India (2002)
3. Lemaignan, C., Motta, A.T.: Zirconium alloys in nuclear applications. Reprint from Materials Science and Technology, A Comprehensive Treatment/edited by RW Cahn, PH Kramer, Chapter 7, from V. 10 B, Nuclear Materials, Part 2, Edited by BR (1993)
4. Cheadle, B.A., Coleman, C.E., Licht, H.: CANDU-PHWR pressure tubes: their manufacture, inspection, and properties. Nucl. Technol.. Technol. **57**(3), 413–425 (1982)
5. Singh, R.N., Mukherjee, S., Gupta, A., Banerjee, S.: Terminal solid solubility of hydrogen in Zr-alloy pressure tube materials. J. Alloy. Compd. **389**(1–2), 102–112 (2005)

6. Pan, Z.L., Ritchie, I.G., Puls, M.P.: The terminal solid solubility of hydrogen and deuterium in Zr-2.5 Nb alloys. J. Nucl. Mater.Nucl. Mater. **228**(2), 227–237 (1996)
7. Northwood, D.O., Kosasih, U.: Hydrides and delayed hydrogen cracking in zirconium and its alloys. Int. Metals Rev. **28**(1), 92–121 (1983)
8. Goldak, J., Lloyd, L.T., Barret, C.S.: Phys. Rev. **144**, 478 (1966)
9. Bradbrook, J.S., Lorimer, G.W., Ridley, N.: J. Nucl. Mater.Nucl. Mater. **42**, 172 (1972)
10. Carpenter, G.J.C.: J. Nucl. Mater.Nucl. Mater. **48**, 264 (1973)
11. Yamanaka, S., et al.: J. Alloys Compd. **23**, 293–295 (1999)
12. Singh, R.N., Ståhle, P., Massih, A.R., Shmakov, A.A.: Temperature dependence of misfit strains of delta-hydrides of zirconium. J. Alloy. Compd. **436**, 150–154 (2007)
13. Singh, R.N., Ståhle, P., Banks-Sills, L., Ristmanaa, M., Banerjee, S.: δ-hydride habit plane determination in α-Zirconium at 298 K by strain energy minimization technique. Defect and Diffusion Forum **279**(2008), 105–110 (2008)
14. Sharma, R.K., Tewari, A., Singh, R.N., Kashyap, B.P.: Optimum shape and orientation of δ-hydride precipitate in α-zirconium matrix for different temperatures. J. Alloy. Compd. **742**(2018), 804–813 (2018)
15. Sunil, S., Gopalan, A., Bind, A.K., Sharma, R.K., Murty, T.N., Singh, R.N.: Effect of radial hydride on delayed hydride cracking behaviour of Zr-2.5Nb pressure tube material. J. Nucl. Mater.Nucl. Mater. **542**(2020), 152457 (2020)
16. Narayana Murty, T., Singh, R.N., Ståhle, P.: Delayed hydride cracking of Zr-2.5%Nb pressure tube material due to partially constrained precipitates. J. Nucl. Mater.Nucl. Mater. **513**(2019), 129–142 (2019)
17. Singh, R.N., Ståhle, P., Chakravartty, J.K., Shmakov, A.A.: Threshold stress intensity factor for Delayed hydride cracking in Zr-2.5%Nb pressure tube alloy. Mater. Sci. Eng. A **523**(2009), 112–117 (2009)
18. Singh, R.N., Bind, A.K., Srinivasan, N.S., Ståhle, P.: Influence of hydrogen content on fracture toughness of CWSR Zr-2.5Nb pressure tube alloy. J. Nucl. Mater.Nucl. Mater. **432**(2013), 87–93 (2013)
19. R. N. Singh and S. Banerjee (as Co-Authors from BARC) in the Final report of a coordinated research project 1998–2002: Delayed hydride cracking in zirconium alloys in pressure tube nuclear reactors. IAEA, Vienna, 2004, IAEA-TECDOC-1410, pp. 1–92 (2004)
20. ASTM E1820–Standard Test Method for Measurement of fracture toughness, ASTM Book of Standards Volume: 03.01 (2015)
21. Dubey, J.S., Wadekar, S.L., Singh, R.N., Singh, T.K., Chakravartty, J.K.: Assessment of hydrogen embrittlement of zircaloy-2 pressure tubes using unloading compliance and load normalization techniques for determining J-R curves. J. Nucl. Mater. Nucl. Mater. **264**, 20–28 (1999)
22. Bind, A.K., Singh, R.N., Saurav Sunil, J., Chattopadhyay, D.S.: Effect of deuterium content on fracture toughness of Zr-2.5 Nb pressure tube material in the temperature range of ambient to 300 °C. J. Nucl. Mater. Nucl. Mater. **496**(2017), 182–192 (2017). https://doi.org/10.1016/j.jnucmat.2017.08.040
23. Sharma, R.K., Bind, A.K., Avinash, G., Singh, R.N., Tewari, A., Kashyap, B.P.: Effect of radial hydride fraction on fracture toughness of CWSR Zr-2.5% Nb pressure tube material between ambient and 300 °C temperatures. J. Nucl. Mater. Nucl. Mater. **508**(2018), 546–555 (2018)
24. Singh, R.N.: Flow behavior and hydrogen embrittlement of Zr-2.5 wt.% Nb pressure tube alloy. Ph. D. Thesis, IIT Bombay, India (2002)
25. Singh, R.N., et al.: Hydrogen charging, hydrogen content analysis and metallographic examination of hydride in Zirconium alloys. BARC report No. BARC/2003/E/034, pp. 1–45 (2003)

Investigating Deep Learning-Based Stress Prediction in Particulate Polymer Composites Using Multiple Quality Measures

Sristi Gupta[1(✉)], Tanmoy Mukhopadhyay[2], Divyesh Varade[1], and Vinod Kushvaha[3]

[1] Department of Civil Engineering, Indian Institute of Technology Jammu, Jammu, India
sristi.gupta@iitjammu.ac.in

[2] Department of Aeronautics and Astronautics, University of Southampton, Southampton, UK

[3] Independent Researcher, Boise, USA

Abstract. The essential qualities for materials in various engineering and manufacturing industries require minimizing manufacturing costs and ensuring optimal design to reinforce safety and reliability. Particulate polymer composites (PPC) find wider applicability due to their versatility and exceptional structural and mechanical properties. To reduce the computational runtime of finite element modelling (FEM) software typically used for analysis of material mechanical properties, deep learning approaches have gained significant popularity. However, the stress fields generated by these approaches are inefficiently matched with FEM results, particularly when based only on a few parameters. In this study, a pix2pix conditional Generative Adversarial Network (cGAN) is used for predicting the stresses as 2D images in PPC corresponding to the results of ABAQUS CAE 2020 FEM simulations. The results are evaluated corresponding to several image matching quality metrics including Structural Similarity Index Measure (SSIM), Peak Signal to Noise Ratio (PSNR), Correlation coefficient, and Mean Square Error (MSE). In effect, these metrices reveal the deep learning model capabilities to imitate features and subsequently determine the optimal configuration of the model. The experimental results showed strong similarity based on these metrices between the modelled results and the FEM simulation results.

Keywords: Particulate Polymer Composites · Deep Learning · Stress fields · ABAQUS · Finite element analysis

1 Introduction

Analyzing the material mechanical properties is a fundamental process in manufacturing and fabrication. Finite Element Modeling (FEM) is a computational scheme employed for predicting the mechanical behavior of different materials, although it frequently necessitates substantial computational resources and time [1]. However, Finite Element Analysis (FEA) continues to be crucial and widely employed method for analyzing the mechanical response in materials, due to its capability to yield highly informative

A. Tiwari et al. (Eds.): iConFIRM 2023, SPPHY 308, pp. 101–107, 2024.
https://doi.org/10.1007/978-981-97-4557-9_11

outcomes that align with validated experimental results [2]. Subsequently, FEM simulations have been extensively accepted and used in assessing the mechanical behavior of polymer composites. While traditional Finite Element Method (FEM) simulations have long been used to foster understanding, Deep Learning (DL) approaches are emerging as promising tools for predicting the mechanical properties of Particulate Polymer Composite (PPC) materials [1, 3].

The DL models, specifically conditional Generative Adversarial Networks (cGANs), have demonstrated significant capabilities to learn complex patterns and predict accurate stress field predictions in 2D images [4, 5]. By training these models with simulated mechanical properties data from FEM software for polymer composites subjected to various types of loading, a stress distribution pattern can be imitated. Thus, faster and efficient predictions of stress distribution patterns in polymer composites can be achieved. All in all, to comprehend the stress patterns effectively, utilization of AI tools is crucial. Subsequently, the evaluation of accuracy received from AI-based algorithm necessitates the application of certain quality assessment methodologies [6].

The results of the DL algorithm can be compared visually with the real data obtained from the FEA simulations. However, it can be challenging to evaluate large datasets based on visual interpretation. Further, this process is not only highly time-consuming but a taxing job. To ascertain the comparison between the real and predicted data, popular quality metrics are employed, namely, Structural Similarity Index Measure (SSIM), Peak Signal-to-Noise Ratio (PSNR), Mean Squared Error (MSE), and Least Square Error (L2 Norm) [7–9]. These metrics provide a quantitative measure on the performance of the deep learning model which gives a comprehensive assessment of the model's effectiveness. This study aims to go beyond the single parameter evaluation approach and investigates multiple image-matching quality measures for a more comprehensive evaluation [10, 11].

2 Materials and Methods

2.1 FEA Simulations

The validation of the experimental and computational results for the short fiber reinforced polymer composite is imperative for several reasons. To streamline the complexity of the problem, Finite Element Analysis (FEA) simulations are performed on a two-dimensional (2-D) Representative Area Element (RAE) of $600 \times 600\ \mu m^2$. Notably, this approach reveals a robust validation with experimentally tested dog-bone-shaped specimens when compared to the computational results [12].

The results obtained from the Finite Element Analysis (FEA) simulations play a crucial role in generating training data for the AI-based algorithm. In this process, the AI model utilizes Young's modulus values for each material, systematically adjusting the mechanical properties within a range of $\pm 30\%$ [5]. The corresponding images associated with these varied Young's modulus values are entered into the AI algorithm. In addition to Young's modulus of each material, the third feature involves capturing time steps representing frames in the FEA results. These extracted parameters provide the AI algorithm with the ability to discern stress patterns effectively. The output generated by the AI-based algorithm presents stress patterns in each material, delineating the maximum and minimum ranges of stresses, thereby ensuring a meaningful interpretation of the data.

2.2 Pix2pix cGAN Stress Analysis

In this study, to train the AI model effectively, a dataset is derived from Finite Element Analysis (FEA) simulations. This dataset contains 1536 images illustrating stresses in the x-direction, extracted from ABAQUS CAE 2020 output database (odb) files. Split into training and testing sets, it consists of 256 odb files. Each file yields images across 6 time steps, capturing diverse stress patterns related to different material properties [5]. Hereafter, the labeled training images (8-bit true color images) representing stresses are prepared for the respective extracted data and fed to the AI algorithm (Fig. 1).

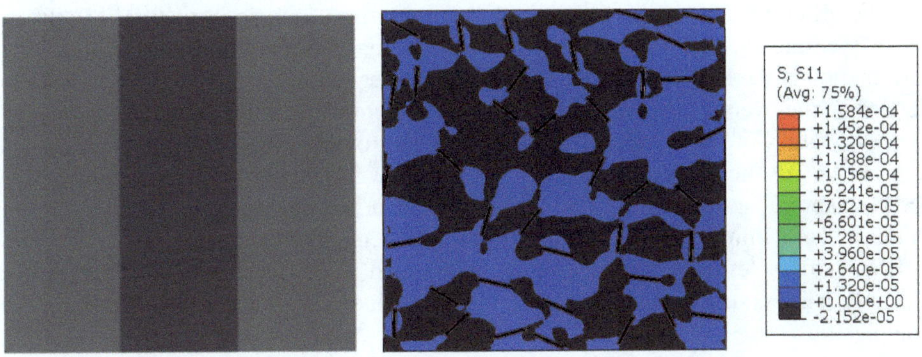

(a) Grayscale input image (b) Maximum S_{11} stresses in epox y

Fig. 1. Grayscale input image (512 x 512) featuring three distinct strips representing material properties (short fibers and epoxy) and time steps referring to image frames of the stresses corresponding in the RAE of PPC in figure (b) for maximum S11 stresses (blue colored) in the epoxy [6] (Color figure online)

Here, we have combined the cGAN algorithm with a finite element-based tool to predict stress fields in short fiber reinforced composites. Pix2pix cGAN model gives the output that is conditioned on the input (source) image. The pix2pix cGAN employs two neural networks: the generator produces realistic images, and the discriminator distinguishes between real and generated images. While the training progresses, the generated image quality is expected to improve; however, additional training epochs may not necessarily lead to further improvement in image quality [13].

Our proposed methodology involves a quantitative evaluation of cGAN results using various image-matching quality metrics, including SSIM, PSNR, MSE, and L2 Norm. Most extensively used image quality metrics such as PSNR, MSE are difference-based measures whereas SSIM is based on the Human Visual System (HVS) [7].

2.3 Evaluation Methods

Notably, evaluating the results of the cGAN model using these metrics helps in gaining a better understanding of the prediction quality of the DL-based pix-2-pix algorithm results. The input dataset prepared for cGAN model consists of grayscale images (Fig. 1a) referring to the maximum and minimum stresses developed in the individual materials fabricating short fiber reinforced polymer composite. The properties of the polymer composite are used as important labels as they affect the behavior of the materials subjected to certain loading conditions. The images belonging to the maximum and minimum stresses developed in the carbon fibers and epoxy are taken as the reference dataset for comparing the accuracy of the AI model. The predicted images processed from the DL-based algorithm are assessed for accuracy using the different similarity metrics using MATLAB version R2023a. The analysis gives an insight into the model's capability to capture the accurate stress pattern and determine the optimal configuration for accurate predictions.

The different similarity metrics used for comparison include: Structural Similarity Index Measure (SSIM) which is based on the primary assumption that the structural information is the key feature in the HVS. SSIM derives three specific components to determine the overall quality: the luminance, the contrast, and the structural similarity [7]. The values of the weights for each of three aforementioned components are set to one. An advancement of the SSIM, Multi-scale Structural Similarity Index (MS-SSIM) takes into account the variations in structure at multiple scales of detail within an image by combining the SSIM value of numerous versions of images at different scales [14] and offers several advantages over SSIM with some disadvantages as well. For example, it is less suitable for real-time applications due to its slow and resource-intensive complexities.

The other metrics used to evaluate the similarity between the results include the mean squared error (MSE), peak signal to noise ratio (PSNR) [8]. MSE, simply measures the error by taking average squared difference between pixels for the real and predicted values, while PSNR determines the lost information in predicted images. PSNR is primarily used for quality assessment of compressed images and videos [15]. For images of 8-bit data type, quality assessment typically revolves around a Signal to Noise Ratio (SNR) of range 30 dB to 50 dB. A higher SNR indicates better image quality, with values above 40 dB signifying high-quality image predictions. Conversely, images with SNR values below 20 dB are deemed unacceptable.

3 Results and Discussions

In order to evaluate the performance of the cGAN-based pix2pix algorithm, several performance metrics are used to evaluate the processed predicted images depicting the stress patterns of short fiber reinforced polymer composite. There are 256 images in the test dataset that are examined for accuracy against the reference dataset. The analysis reveals a strong correlation between the predicted images and original dataset, particularly for both maximum and minimum stress pattern in x-direction for carbon fibers and epoxy, with correlation values approaching value 1. The study employs per-pixel loss metrics, detailed in Fig. 2, to quantify the performance of the algorithm.

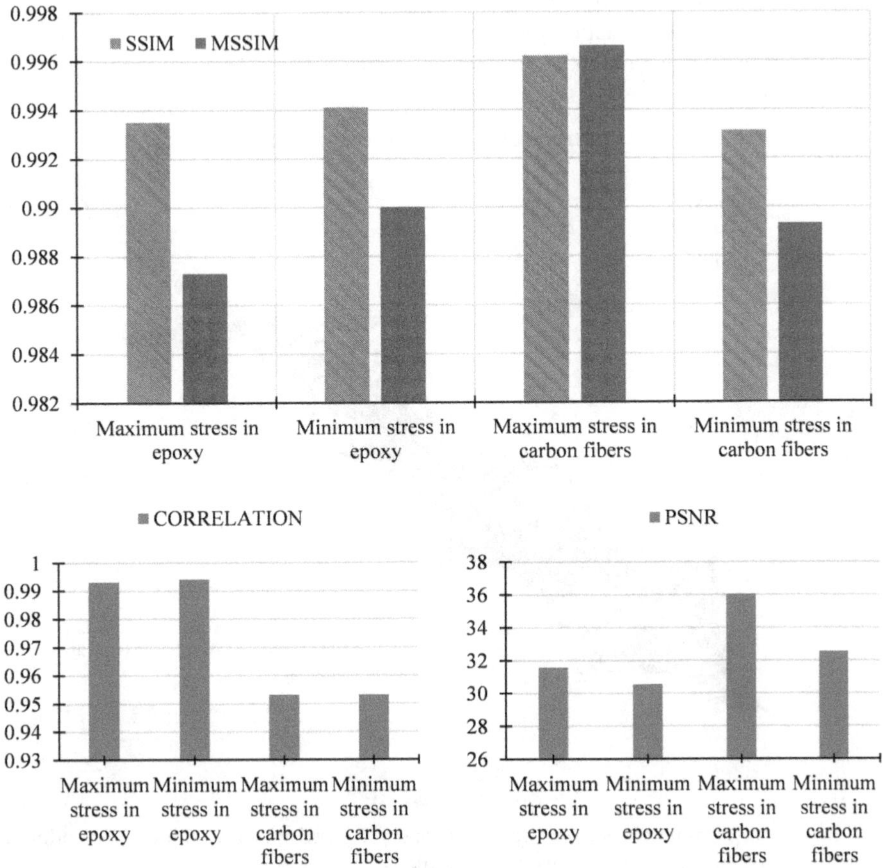

Fig. 2. Quality assessment outcomes showing the similarity between reference and predicted datasets within the testing set, for data obtained for maximum and minimum stress patterns in the x-direction across epoxy and short carbon fiber composites

Quality assessment results for predictions obtained from an AI-based algorithm for maximum and minimum stress patterns in the x-direction across epoxy and short carbon fiber composites show exceptionally good performance for stresses developed in the x-direction, indicating lower complexity and a high correlation near 1 for all cases of maximum and minimum stresses in each material. The stresses in fibers are higher than in epoxy, resulting in a slightly lower correlation numerical value. However, the PSNR value for maximum stresses in carbon fibers is higher than in other cases, indicating a good quality of the reconstructed signal. This suggests that the reconstructed signal closely matches the original signal with minimal noise. Additionally, SSIM and MSSIM values exhibit remarkable closeness between reference and predicted images for stresses in each material in the minor principal direction. In Fig. 3, the coefficient of determination (R^2) and Root Mean Square Error (RMSE) illustrates the agreement between the reference and predicted data. The R^2 value of 0.9848 suggests a strong linear relationship between the predicted and reference data. The statistical agreement emphasizes the effectiveness of the AI approach in achieving accurate predictions of stress patterns [16].

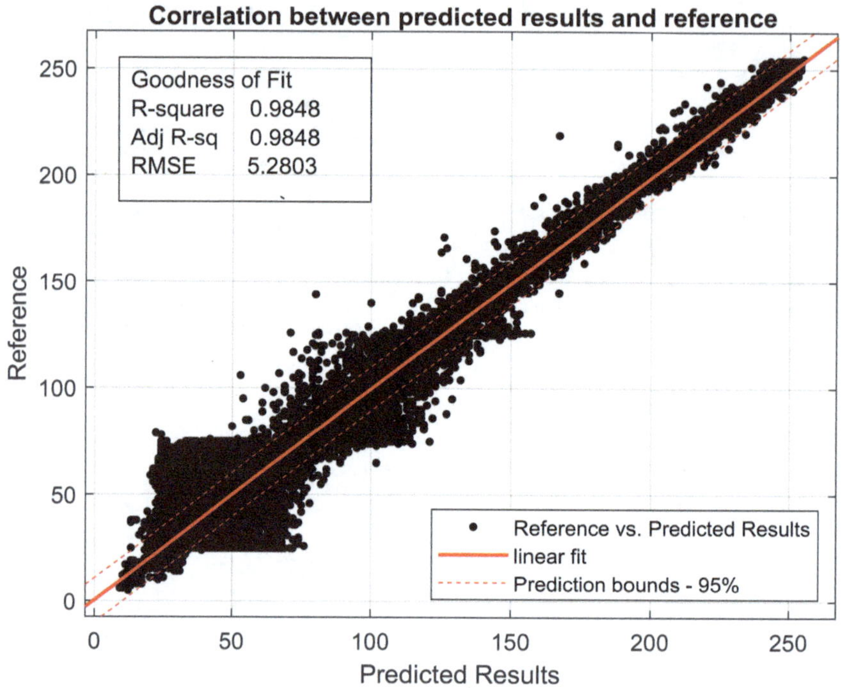

Fig. 3. Pixel-wise correlation between reference and predicted images, with corresponding R^2 and RMSE values indicating the quality of the results

4 Conclusion

Our study highlights the effectiveness of the pix2pix cGAN model in predicting the stress levels in PPC. The evaluation metrics, including SSIM, MSSIM, PSNR, correlation coefficient, and MSE, provide valuable insights into the model's performance. These insights can guide the optimization of the cGAN architecture, enabling researchers to enhance the model's accuracy and reliability in predicting stress levels. The research findings present possibilities for further exploration and development of more accurate and reliable stress prediction models using deep learning techniques.

References

1. Sharma, A., Mukhopadhyay, T., Rangappa, S.M., Kushvaha, V.: Advances in computational intelligence of polymer composite materials: machine learning assisted modeling, analysis and design. Arch. Comput. Meth. Eng. **29**, 3341–3385 (2022)

2. Ahmadi Moghaddam, H., Mertiny, P.: Stochastic finite element analysis framework for modelling mechanical properties of particulate modified polymer composites. Materials (Basel) **12**(17), 2777 (2019). https://doi.org/10.3390/ma12172777.PMID:31470532;PMCID:PMC6747834
3. Bhaduri, A., Gupta, A., Graham-Brady, L.: Stress field prediction in fiber-reinforced composite materials using a deep learning approach. Compos. B Eng. **238**, 109879 (2022)
4. Yang, Z., et al.: Deep learning model to predict complex stress and strain fields in hierarchical composites. Sci. Adv. **7** (2021). https://doi.org/10.1126/sciadv.abd7416
5. Gupta, S., Mukhopadhyay, T., Kushvaha, V.: Microstructural image based convolutional neural networks for efficient prediction of full-field stress maps in short fiber polymer composites. Defence Technol. **262**, 619 (2022)
6. Kumar, B., Kumar, S.B., Kumar, C.: Development of improved SSIM quality index for compressed medical images. In: 2013 IEEE Second International Conference on Image Information Processing (ICIIP-2013), pp. 251–255 (2013). https://doi.org/10.1109/ICIIP.2013.6707593
7. Wang, Z., Bovik, A.C., Sheikh, H.R., Simoncelli, E.P.: Image quality assessment. From error visibility to structural similarity. IEEE Trans. Image Process. Publ. IEEE Signal Process. Soc. **13**, 600–612 (2004)
8. Winkler, S., Mohandas, P.: The evolution of video quality measurement. From PSNR to hybrid metrics. IEEE Trans. Broadcast. **54**, 660–668 (2008)
9. Wang, J., Chen, P., Zheng, N., Chen, B., Principe, J.C., Wang, F.-Y.: Associations between MSE and SSIM as cost functions in linear decomposition with application to bit allocation for sparse coding. Neurocomputing **422**, 139–149 (2021)
10. Kotevski, Z., Mitrevski, P.: Experimental comparison of PSNR and SSIM metrics for video quality estimation. In: Davcev, D., Gómez, J.M. (eds.) ICT Innovations 2009, pp. 357–366. Springer, Heidelberg (2010). https://doi.org/10.1007/978-3-642-10781-8_37
11. Horé, A., Ziou, D.: Image quality metrics: PSNR vs. SSIM. In: 2010 20th International Conference on Pattern Recognition, Istanbul, Turkey, pp. 2366–2369 (2010). https://doi.org/10.1109/ICPR.2010.579
12. Kushvaha, V.: Dissertation on Synthesis, Processing and Dynamic Fracture Behavior of Particulate Epoxy Composites with Conventional and Hierarchical Micro-/Nano-fillers. Auburn University, Alabama (2016)
13. Abdelmotaal, H., Abdou, A.A., Omar, A.F., El-Sebaity, D.M., Abdelazeem, K.: Pix2pix conditional generative adversarial networks for scheimpflug camera color-coded corneal tomography image generation. Transl. Vis. Sci. Technol. **10**, 21 (2021)
14. Wang, Z., Simoncelli, E.P., Bovik, A.C.: Multiscale structural similarity for image quality assessment. In: The Thrity-Seventh Asilomar Conference on Signals, Systems & Computers, Pacific Grove, CA, USA, vol. 2, pp. 1398–1402 (2003). https://doi.org/10.1109/ACSSC.2003.1292216
15. Thakur, N., Devi, S.: A new method for color image quality assessment. Int. J. Comput. Appl. **15**, 10–17 (2011)
16. Tang, K., et al.: Machine learning-based morphological and mechanical prediction of kirigami-inspired active composites. Int. J. Mech. Sci. **266**, 108956 (2024). https://doi.org/10.1016/j.ijmecsci.2023.108956

Synthesis and Characterization of Ausformed Nanostructured Bainite

Abhinesh Verma(✉) and Khushboo Rakha

Metallurgical and Materials Engineering, Indian Institute of Technology Ropar, Rupnagar 140001, Punjab, India
krakha@iitrpr.ac.in

Abstract. The ausforming process of plastically deforming the austenite and then isothermally holding it at a bainitic temperature helps a lot in increasing the transformation kinetics and tailoring the microstructure and mechanical properties of the material. This study aims to evaluate the influence of thermomechanical treatment, followed by isothermal treatment on the mechanical properties and phase distribution characteristics. The optical and scanning electron microscopy of the bainite will be used to analyze the effects. The process involves the plastic deformation of austenite at a temperature of approximately near to the industrial coiling temperature followed by isothermal holding in the bainitic temperature range. A long isothermal holding time was used to allow the complete transformation of the plastically deformed parent austenite phase into the final nanostructured bainitic morphology consisting of bainitic ferrite with film austenite. The final microstructure obtained after long isothermal heating was studied and correlated with the hardness of the specimen thus obtained by applying different ausforming strains. Results showed that the arrangement of microstructural constituents leads to a decrease in the hardness.

Keywords: Nanostructured Bainite · Ausforming · Retained Austenite · Deformation temperature · Ausforming strain

1 Introduction

The invention of nanostructured bainite has given a very potential candidate to the industry to be used in a variety of applications where a combination of properties is required. Decreasing the grain size to the nano level helps in achieving a very high strength. The microstructure consists of very fine acicular ferrite and film-type retained austenite [1]. Getting a microstructure consisting of acicular ferrite and film austenite gives it a remarkable set of properties that helps in achieving high hardness(600-700 HV), ultra-high strength (2–2.3 GPa), and toughness of 30MPam$^{(-1/2)}$ [2–6]. Changing the composition and isothermal transformation temperature can efficaciously help in controlling the microstructure [7]. When the material is isothermally transformed at a low-temperature range of 200–400 °C, parent phase austenite is consumed into very thin size ferrite and film-shaped austenite with a width of the order of a few tens of nanometres. Also, the average thickness of ferrite plates and austenite films was found to decrease

A. Tiwari et al. (Eds.): iConFIRM 2023, SPPHY 308, pp. 108–116, 2024.
https://doi.org/10.1007/978-981-97-4557-9_12

with the decrease in the austempering temperature [8]. The displacive transformation nature of the nanostructured bainite has been shown in one of the work with the help of atomic scale characterization [9]. Nano structured bainite with 2.3 GPa, with 30% ductility and 40 MPa/m$^{(1/2)}$ has been achieved at lab scale [10]. This is only possible by the addition of carbon inside steel which helps in achieving the transformation at low temperatures. Silicon is the major alloying element added to suppress the carbide precipitation by lowering the martensite start temperature [11]. Although, the carbide precipitates has still been found in few steels even after the addition of 1.5% silicon in the steel [12]. Carbide-free bainitic steels has also been developed only with a combination of ferrite and austenite which have good ductility because of the presence of film-type austenite in it [12].

High strength of nano bainitic steels is mainly obtained due to impediment of dislocation motions due to the presence of large number of interface boundaries at nano meter distance. This interface density increases with increase in driving force for transformation i.e., decrease in the transformation temperature. Ideally decreasing the grain size decreases the ductility but the presence of film type retained austenite helps in retaining the material to be ductile [13]. Although, a separate population of austenite also exists in this material in blocky form which is a low carbon form of austenite and less stable and may undergo martensitic transformation under loading conditions giving TRIP effect [14]. Although the properties like strength and ductility can't be generalized as they depend on the working conditions. So, application and loading-based testing should be done to interpret the behavior of material in actual working conditions as there could be many design factors like splines, threads, holes, casting inclusions inside the component.

Transformations at such a low temperature leads to very slow atomic diffusion resulting in very long transformation times in case of big components [2]. It is evident that a prolonged transformation process will result in higher processing costs for the component, unless it is intended for a specific use where the costs can be justified. So, in this regard researchers have tried to increase the transformation kinetics by using certain alloying elements like aluminum and cobalt [15]. Both of those elements have been found helping in increasing the reaction kinetics and retaining the austenite within the microstructure at room temperature. In contrast to this, it has been observed in one of the research that the addition of aluminum beyond a certain atomic percent start deteriorating the properties and increasing the martensite start temperature of the steel, so addition beyond a certain level did not helped much in this [16]. Multi step austempering has also been found increasing the reaction kinetics in some of the studies in low carbon steels. This helped in consuming the big austenite blocks and two populations of the bainite were formed with much finer microstructure [17].

Additionally, one other important factor i.e., ausforming has been used by the researchers to explore its effect on transformation kinetics. It has been found that the transformation was accelerated by the application of strain on the steel [18–20]. The idea behind using thermo-mechanical deformation is that it will introduce dislocations inside material which will act as a new potential site for nucleation of new bainitic nuclei. Ausforming temperatures and strains are crucial factors in determining the final mechanical properties of a material based on its microstructure. A lot of work has been done in this direction at the lab scale. The commercialization of making nanostructured bainite by ausforming and isothermal holding is the next step to be executed.

In this article, Lab-scale casting and thermomechanical treatment were followed by isothermal holding to facilitate translation to an industrial-scale rolling mill. The current simulation was carried out to understand its effect on the microstructural and mechanical properties.

2 Experimental Procedure

The chemical compositions used for research the current work was selected from literature as Fe-0.79C-1.49Si-1.95Mn-0.97Cr-0.24Mo-1.60Co-0.99Al(wt.%) which has shown the enhanced kinetics on the basis of thermodynamics calculations [4]. A flat slab of 40 mm thickness was obtained post casting. Small cylindrical blocks of diameter 10 mm and length 15 mm were machines from the hot rolled specimen. Further, samples were homogenized at 1250 °C for 8 h packed in a tube in argon environment. Samples were austenitized at a temperature of 1000 °C for 180 s followed by ausforming strain of 7%, 15%, 30% and 60% at 570 °C along with isothermal holding at 350 °C for 10 days. The schedule used for thermomechanical treatment has been shown in Fig. 1.

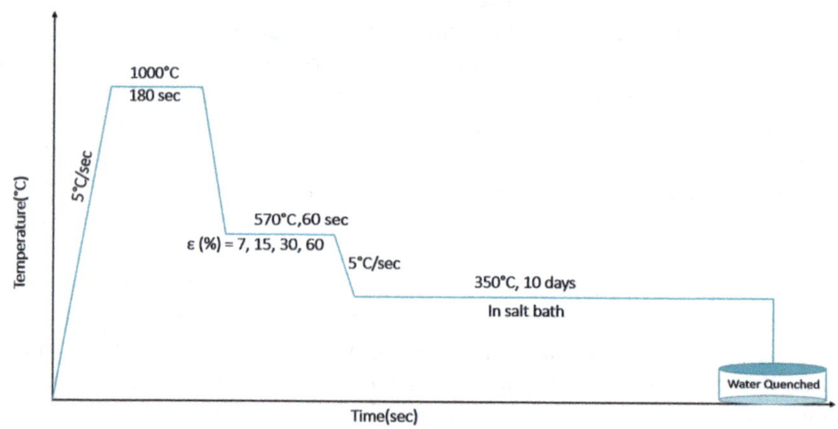

Fig. 1. Thermomechanical schedule

A servo test thermomechanical treatment simulator (TMTS) of capacity 500kN along with rapid salt bath quencher was used to study the effect of deformation on microstructure/morphology of phases transformed. For isothermal holding of specimen salt bath quencher was used for long duration holding followed by water quenching.

Figure 2 shows the specimen bisected along longitudinal axis and the arrows represents the loading direction. Samples were polished with the conventional techniques using abrasive papers from coarser to finer grit size as P200, P400, P600, P800, P1000, P1200, P1500 followed by fine diamond paste of 0.5 μm particle size. 2% Nital solution was used for etching to make bainitic sheaves clearly visible. FALCON 603 type Vickers hardness tester was used to measure the hardness of the specimen along the midway

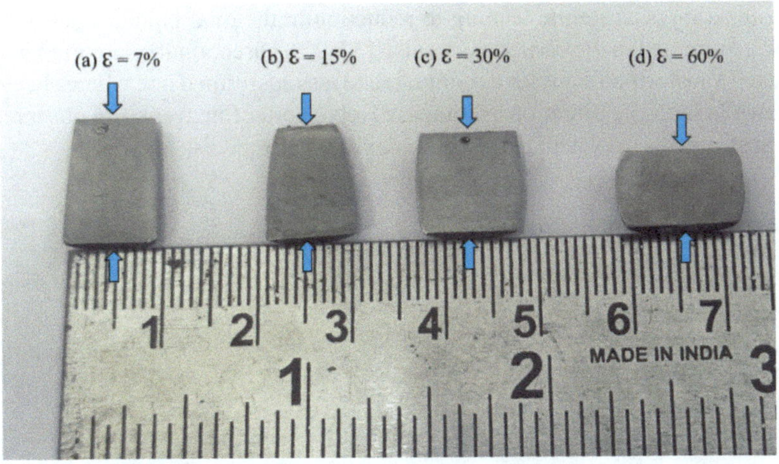

Fig. 2. Images of ausformed samples

portion of the bisected specimen. The hardness values were accurately measured on a scale of HV10 (10 kgf) for a dwell time of 15 s, extracted from the mid-way of vertically sectioned samples. Microstructural characterization was carried out by standard procedures, using a Leica optical microscope and JEOL 6610 V electron microscope, after appropriate mirror finishing.

3 Results and Discussion

Figure 3 represents the optical micrograph taken for specimen ausformed at different compression strains of 7,15,30 and 60%. A fully bainitic structure consisting of alternate layers of austenite and ferrite could be observed in the optical images after the thermomechanical treatment of samples. The black needle shapes in the images can be termed as bainitic ferrite and the white portion could be assumed as retained austenite or martensite. The optical images clearly show that the bainitic laths are becoming shorter and thinner from long and thicker initially by increasing the strain from 7 to 60%. The austenite is categorized into two categories one in film shape embedded with bainitic ferrite and the other in blocky form present in between the other microstructural constituents. The reason for finer microstructure is attributed to the increase in the ausforming strain which is thinning the bainitic ferrite and film shape retained austenite further. Ausforming, is the process of generating dislocations within austenite grains, which act as nucleation sites for the formation and growth of other nuclei. The large number of small nuclei, along with dislocation debris, prevents the growth of adjacent nuclei, resulting in shorter and finer nuclei [22, 25–27].

The deformation of austenite grains starts transforming into acicular ferrite grains along with the austenite films inside it. But, a few in number austenite blocks were also found in between the different variants formed under stresses. It was observed that with the increase in the ausforming strain more blocks of austenite were observed depicting the over-stabilization of austenite which is ductile in nature. This is a clear indication of over-stabilization of austenite leading to reduction in the final bainite fraction. Similar results were reported in the earlier works also [20–23]. Directional orientation was found in the heavily ausformed samples as compared to less ausformed one where sheaves have been grown in multiple directions. Comparatively shorter Sheaves were observed in the heavily ausformed samples than that of the lightly deformed specimen.

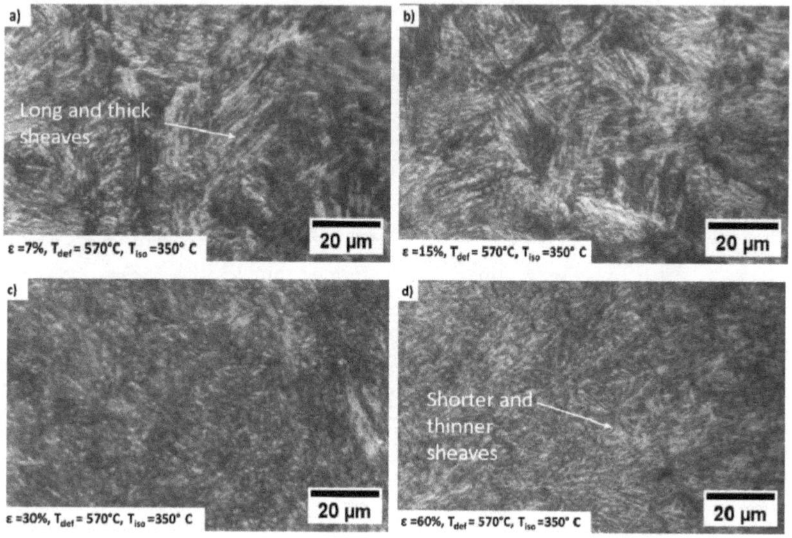

Fig. 3. Optical images of sample ausformed with a)7%, b) 15%, c)30%, and d) 60% strain

Samples were bifurcated along the longitudinal axis using wire EDM so that the mechanical properties of the specimen surface are not disturbed. The Vickers hardness testing along with the change in ausforming strain is shown in the Fig. 5. It can be clearly observed that the compression strain affected the strength of the steel as there occurred a change in the hardness values. The hardness of the four-specimens declined as the strain increased from 7% to 60% during deformation at 570 °C followed by isothermally holding at 350 °C for 10 days. The morphologies in Fig. 4 shows that along with change in microstructure from coarser to finer one, a greater number of blocks of retained austenite also started appearing.

The hardness of the specimen here is found to be decreasing along with the increase in the ausfroming strain. This is because an increase in ausforming strain leads to refinement in the austenite grain size along with the introduction of new dislocations inside it which increases the possible number of nucleation sites for the generation of new ferrite plates. The more the number of dislocations and austenite grain boundary area per unit volume, the more ferrite plates would be thus formed and it will entrap a higher volume fraction of non-transformed retained austenite in film-type and blocky form. This process results in an overall increase in the volume fraction of retained austenite, in both film and blocky form, and consequently leads to a reduction in the hardness of the specimen with the increase in ausforming strain. Ausforming temperature, ausforming strain, and

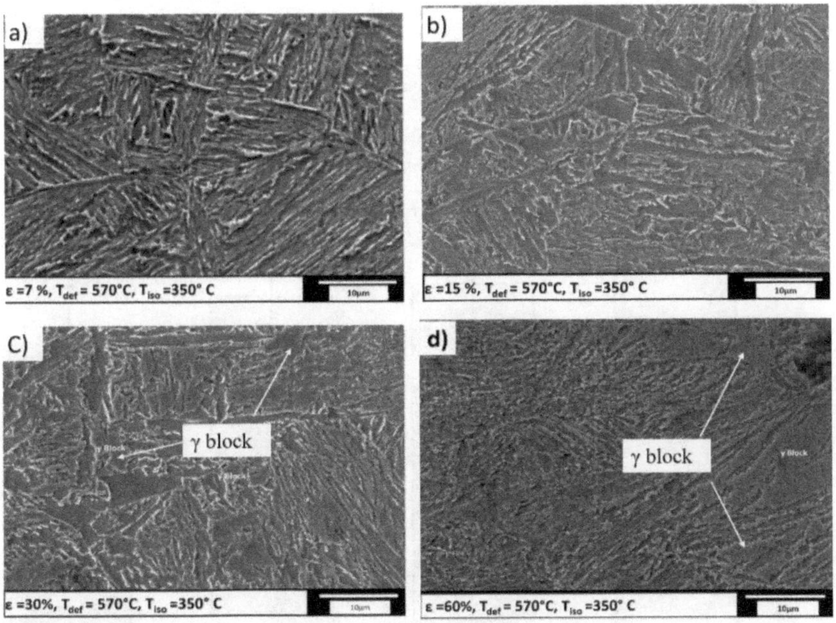

Fig. 4. SEM images of sample ausformed with a) 7%, b) 15%, c) 30%, and d) 60% strain

strain rate are the important factors that mainly affect the change in hardness of the specimen. T. Kumnorkaew et al. have explained that at lesser strains higher martensite phase fraction is the reason for the high hardness values whereas at higher strain values larger retained austenite fraction caused the reduction in hardness. Also, ausforming resulted in austenite grain refinement leading to an increase in the matrix strength which increased the overall retained austenite fraction [24]. S.Golchin et al. have also found that Ausforming have also led to increase in the retained austenite fraction in the matrix [26].

So, detailed studies like x-ray diffraction would be needed to estimate the phase fractions of retained austenite along with bainitic ferrite and further correlate it with the mechanical properties.

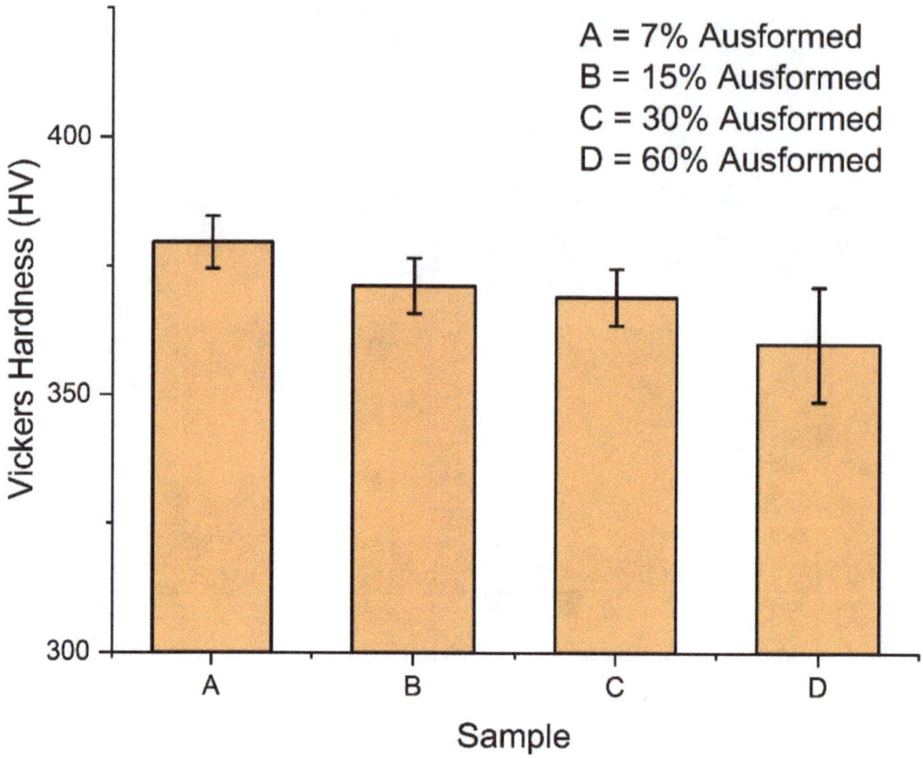

Fig. 5. Vickers hardness graph of 7%, 15%, 30% and 60% ausformed sample.

4 Conclusion

- The microstructure underwent refinement as the ausforming strain increased, leading to a reduction in the prior austenite grain size. Bainitic ferrite sheaves and austenite films have become thinner and shorter from long and slender with the increase in ausforming strain. This indicates a greater transformation in the phase due to elevated driving force.
- Higher ausforming strains resulted in more strengthened ausformed matrix leading to more retained austenite films in ferrite matrix, increasing the overall austenite fraction along with the formation of thicker blocks of austenite, indicating the overstabilization of austenite under substantial strains.
- The reduction in hardness is attributed to the rise in overall austenite fraction filmy as well as blocky austenite within the specimen, as revealed by the electron microscopy images. However, a more comprehensive analysis of phase fractions using Rietveld analysis in X-ray diffraction will be conducted to precisely determine the fraction of phases in bulk specimens.

References

1. Caballero, F.G., Garcia-Mateo, C., Miller, M.K.: Design of novel bainitic steels: moving from ultrafine to nanoscale structures. Jom **66**(5), 747–755 (2014)
2. Bhadeshia, H.K.D.H.: Nanostructured bainite. In: Proceedings of the Royal Society A: Mathematical, Physical and Engineering Sciences, vol. 466, no. 2113. Royal Society, pp. 3–18 (2010)
3. Avishan, B., Garcia-Mateo, C., Morales-Rivas, L., Yazdani, S., Caballero, F.G.: Strengthening and mechanical stability mechanisms in nanostructured bainite. J. Mater. Sci. **48**(18), 6121–6132 (2013)
4. Caballero, F.G.: Ultra-high-strength bainitic steels. ISIJ Int. **45**(11), 1736–1740 (2005)
5. García-Mateo, C., Caballero, F.G., Bhadeshia, H.K.D.H.: Mechanical properties of low-temperature bainite. Mater. Sci. Forum **500–501**, 495–502 (2005)
6. Caballero, F.G., Bhadeshia, H.K.D.H.: Development of Hard Bainite. ISIJ Int. **43**(8), 1238–1243 (2003)
7. Kumar, A., Singh, A.: Toughness dependence of nano-bainite on phase fraction and morphology. Mater. Sci. Eng. A **729**, 439–443 (2018)
8. Lee, S.I., et al.: Influence of austempering temperature on microstructure and mechanical properties of high-carbon nanostructured bainitic steels. Mater. Sci. Eng. A **848**,(2022)
9. Caballero, F.G., Miller, M.K., Garcia-Mateo, C., Cornide, J.: New experimental evidence of the diffusionless transformation nature of bainite. J. Alloys Compd. **577**(Supl. 1), S626–S630 (2013)
10. Sourmail, T., Smanio, V.: Low temperature kinetics of bainite formation in high carbon steels. Acta Mater. **61**(7), 2639–2648 (2013)
11. Kozeschnik, E., Bhadeshia, H.K.D.H.: Influence of silicon on cementite precipitation in steels. Mater. Sci. Technol. **24**(3), 343–347 (2008)
12. Rakha, K., et al.: On low temperature bainite transformation characteristics using in-situ neutron diffraction and atom probe tomography. Mater. Sci. Eng. A **589**, 303–309 (2014)
13. Królicka, A., Żak, A.M., Caballero, F.G.: Enhancing technological prospect of nanostructured bainitic steels by the control of thermal stability of austenite. Mater. Des. **211**, 110143 (2021)
14. Singh, K., Kumar, A., Singh, A.: Effect of prior austenite grain size on the morphology of nano-bainitic steels. Metall. Mater. Trans. A **49**(4), 1348–1354 (2018)
15. Garcia-Mateo, C., Caballero, F.G., Bhadeshia, H.K.D.H.: Acceleration of low-temperature bainite. ISIJ Int. **43**(11), 1821–1825 (2003)
16. Akram, M., Soliman, M., Palkowski, H.: Nano-bainitic steels: acceleration of transformation by high aluminium addition and its effect on their mechanical properties. Metals (Basel). **11**(8) (2021)
17. Avishan, B., Tavakolian, M., Yazdani, S.: Two-step austempering of high performance steel with nanoscale microstructure. Mater. Sci. Eng. A **693**, 178–185 (2017)
18. Gong, W., Tomota, Y., Koo, M.S., Adachi, Y.: Effect of ausforming on nanobainite steel. Scr. Mater. **63**(8), 819–822 (2010)
19. Guo, H., Feng, X., Zhao, A., Li, Q., Chai, M.: Effects of ausforming temperature on bainite transformation kinetics, microstructures and mechanical properties in ultra-fine bainitic steel. J. Mater. Res. Technol. **9**(2), 1593–1605 (2020)
20. Guo, H., Fan, Y.P., Li, Z.X., Li, Q., Feng, X.Y.: Effect of ausforming parameters on strain-induced phase transformation and isothermally transformed bainite. J. Mater. Res. Technol. **11**, 982–991 (2021)
21. He, J., Du, J., Zhang, W., Zhang, C., Yang, Z.G., Chen, H.: abnormal anisotropic dilatation during bainitic transformation of ausformed austenite. Metall. Mater. Trans. A **50**(2), 540–546 (2019)

22. Zhao, J., et al.: Transformation behavior and microstructure feature of large strain ausformed low-temperature bainite in a medium C - Si rich alloy steel. Mater. Sci. Eng. A **682**, 527–534 (2017)
23. Maalekian, M., Kozeschnik, E., Chatterjee, S., Bhadeshia, H.K.D.H.: Mechanical stabilisation of eutectoid steel. Mater. Sci. Technol. **23**(5), 610–612 (2007)
24. Kumnorkaew, T., Lian, J., Uthaisangsuk, V., Bleck, W.: Effect of ausforming on microstructure and hardness characteristics of bainitic steel. J. Market. Res. **9**(6), 13365–13374 (2020)
25. Guo, H., Feng, X., Zhao, A., Li, Q., Chai, M.: Effects of ausforming temperature on bainite transformation kinetics, microstructures and mechanical properties in ultra-fine bainitic steel. J. Market. Res. **9**(2), 1593–1605 (2020)
26. Bhadeshia, H.K.D.H., Matsuzaki, A.: Effect of austenite grain size and bainite morphology on overall kinetics of bainite transformation in steels. Mater. Sci. Technol. **15**(5), 518–522 (1999)
27. Jin, X.J., Min, N., Zheng, K.Y., Hsu, T.Y., Zuyao, X.: The effect of austenite deformation on bainite formation in an alloyed eutectoid steel. Mater. Sci. Eng. A, 438–440, 170–172 (2006)
28. Golchin, S., Avishan, B., Yazdani, S.: Effect of 10% ausforming on impact toughness of nano bainite austempered at 300°C. Mater. Sci. Eng. A **656**, 94–101 (2016)

Vacuum Brazing of Tungsten and Fe-Co-Ni Alloy Along with Micro-Structural Evaluation of Joint Interface Using SEM

Abhishek Upadhyay[1(✉)], Sandeep Kumar[1], Anubhaw Kumar Singh[1], Pal Dinesh Kumar[1], and Neha Sardana[2]

[1] Terminal Ballistics Research Laboratory, Sector 30, Chandigarh 160030, India
abhiups007@gmail.com
[2] Indian Institute of Technology Ropar, Rupnagar 140001, India

Abstract. In the present work, Fe-Co-Ni alloy and tungsten samples (sheet form) have been brazed using Oxygen-Free Electronic (OFE) copper (foil form) as brazing alloy under a vacuum of about 10^{-5} mbar and the interface has been characterized using scanning electron microscopy (SEM) & energy dispersive spectroscopy (EDS). To ensure proper alignment and uniform flow of brazing filler, uniform compressive force was applied using 25 g weight during brazing process. The joint interface was examined using SEM and is found to be thin, continuous and free from micro-cracks. The joint thickness varies between 25 μm to 40 μm. Further, compositional study using EDS along with Dot Mapping & Line Scan has been carried out across the joint interface and analyzed to correlate the diffusion of various elements across the joint. Micro-hardness variations across brazed interface have also been evaluated and the hardness was found to be lowest in filler region than parent materials which indicates the soundness of interfacial joint using this process. The joint strength has been assessed using Ram Tensile Test and the mean value of joint strength has been found to be 91.7 MPa.

Keywords: Vacuum brazing · Capillary action · Dot mapping · Line scan · Micro-hardness

1 Introduction

Vacuum brazing has attained great attention in the recent decades pertaining to their ability to remarkably join dissimilar as well as similar materials combinations [1]. This particular advantage of vacuum brazing has been source of constant motivation and concern for re-searchers, who have to make use of such processes while considering the manufacturing aspects for material combinations that widely differ in physical and thermo-mechanical properties [2]. Vacuum brazing of metal-metal and metal-ceramic combinations have been sought for both critical as well as highly incompatible combinations, particularly materials like W, Mo, alloy steels etc. that are prone to oxidation particularly at high temperatures. Vacuum Brazing, a non-conventional process, involves joining of two parent materials (similar or dissimilar) kept in close contact separated

A. Tiwari et al. (Eds.): iConFIRM 2023, SPPHY 308, pp. 117–123, 2024.
https://doi.org/10.1007/978-981-97-4557-9_13

by a suitable filler alloy, through a carefully designed brazing cycle pattern, along with optimized brazing load (dead weight) in high vacuum environment to avoid oxidation. The brazing filler alloy gets melted at brazing temperature and spreads throughout the joint driven by capillary action and causing wetting of base materials [3, 4]. This filler alloy gets solidified during cooling and forms bonding with the two parent materials. For a successful joint, various factors need to be considered and catered meticulously, that include selection of optimum brazing temperature & soaking time along with the complete brazing cycle pattern, choice of suitable filler alloy and the usage of with/without dead weights at brazing temperature. Eventually the filler gets solidified forming brazed joint with the two parent materials. The present work involves high vacuum brazing of Fe-Co-Ni alloy with Tungsten using OFE Copper as brazing filler alloy. The choice of materials combination is based upon the application in which Fe-Co-Ni alloy needed for leak tight joints in vacuum feedthroughs and is brazed with a refractory material like tungsten using OFE copper to have enhanced electrical erosion resistance due to tungsten being brazed over Fe-Co-Ni alloy.

2 Experimental Work

Samples of Fe-Co-Ni alloy, tungsten were sectioned from available rod and plate respectively and polished initially to achieve surface finish of Rmax < 5 microns, to enable better brazing surfaces. OFE copper washer, fabricated from foil of 100 μm thickness, was used as brazing filler. All three material samples were thoroughly cleaned using ultrasonic solution and subjected to degassing in high vacuum atmosphere at 600°C, to remove various entrapped gases. Subsequently, these samples were placed inside the vacuum brazing furnace as shown in Fig. 1(a) for undergoing high vacuum brazing process (vacuum level of about 10^{-5} mbar). Further, uniform compressive force was applied using 25 g weight to ensure proper alignment of joining components and uniform flow of brazing filler during brazing process. The typical technological curve (not to scale) for brazing of Fe-Co-Ni alloy and W using OFE Copper has been shown in Fig. 2, in which the peak brazing temperature and holding time at peak brazing temperature is denoted by T_b and th respectively. Generally, the peak brazing temperature is about 20–30 °C above the liquidus temperature of filler alloy and the holding time ranges from 5 to 10 min [1, 5]. Thus, value of these parameters that are used in our study are 1100 °C and 5 min respectively. It is further to be noted that the peak brazing temperature, holding time at peak brazing temperature, cooling rate etc. that had been maintained during brazing, were finalised, keeping in consideration the material and thermo-mechanical properties of parent material combinations and brazing filler as mentioned below.

Sample details:

Fe-Co-Ni alloy (cylindrical sample)-Diameter: 15 mm, Height: 10 mm.

Tungsten (sheet): 25 mm × 25 mm × 1 mm.

OFE Copper (washer)-Diameter: 15mm, Thickness: 0.1 mm.

Post-brazing, the brazed assembly obtained was taken out of the hot zone of vacuum furnace and then subjected to visual inspection for any visible defects. Samples were cut and sectioned from the interface area and then subjected to polishing (coarse to fine emery grades), followed by ultrasonic cleaning and then rinsed with isopropyl alcohol

Table 1. Physical and mechanical properties

Property	Density (gm/cc)	Melting Pont (°C)	Young's Modulus (GPa)	Tensile Strength (MPa)	Hardness (HV)	Yield Strength (MPa)	Thermal expansion (10^{-6} K^{-1})
Fe-Co-Ni alloy	8.0	1450	210	520	180	280	5.0
Tungsten	19.3	3410	411	980	430	750	4.5
OFE Cu	8.9	1083	110	220	45	70	16.6

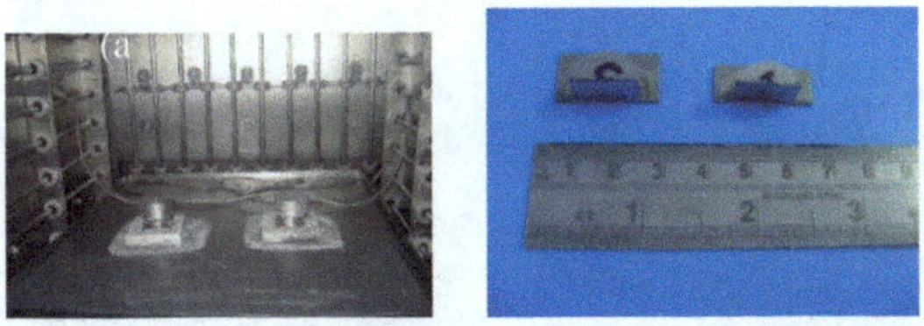

(a) (b)

Fig. 1. Details of samples (a) inside vacuum furnace and (b) cut section of brazed samples

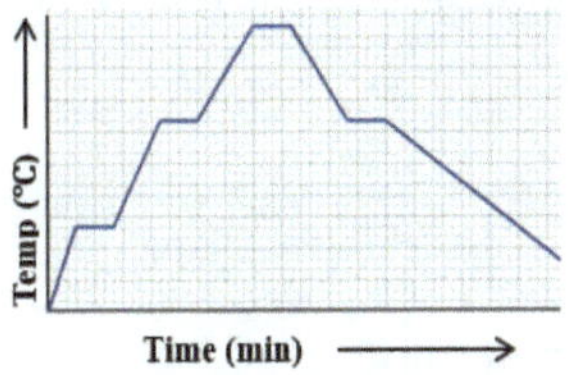

Fig. 2. Typical brazing curve

for SEM examination. SEM/EDS (Line scan & Dot mapping) were used to investigate the Fe-Co-Ni alloy-OFE copper filler-tungsten brazed region (Table 1).

Mechanical strength of the brazed joint has been evaluated using Ram Tensile Test. The Ram tool and fixture plates have been fabricated as depicted by the solid model images in the following Fig. 3. The assembly set up and experimental test set up during testing using Universal Testing Machine (UTM) (Make: FIE) has been shown in Fig. 4(a) and (b) respectively where load has been applied perpendicular to the test set up in compression mode as indicated in Fig. 4 (b).

3 Results and Discussion

SEM (BSE mode) clearly revealed continuous, homogeneous nature of brazed interface. No micro-cracks were observed which is an indication that no brittle compounds/intermetallics [6, 7] have been formed during brazing (Figs. 5 and 6). The brazed joint thickness varied from about 25 μm to 40 μm. (Fig. 6b).

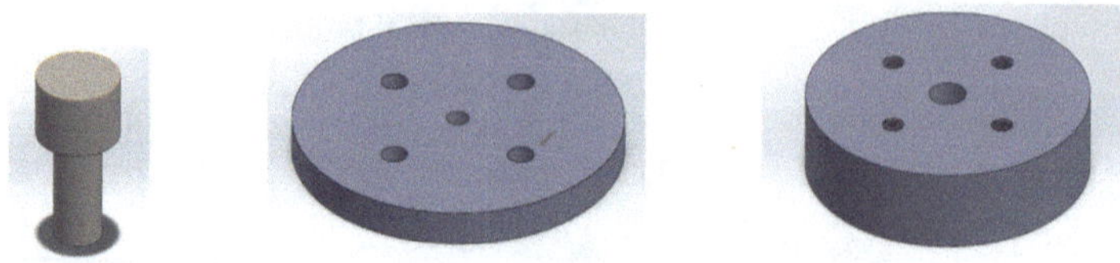

(a) Ram tool (b) Top Plate (c) Base Plate

Fig. 3. Solid models of testing fixture (a) Ram Tool (b) Top Plate (c) Base Plate

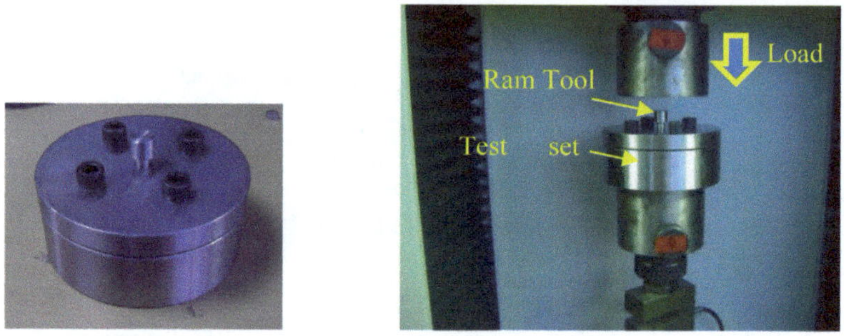

(a) (b)

Fig. 4. (a) Assembly Set Up & (b) Mechanical testing of assembly set up using UTM in compression mode

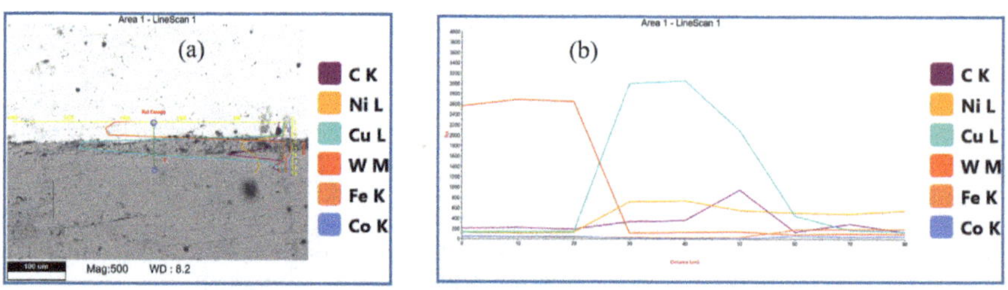

Fig. 5. (a) SEM (BSE Mode) microstructure with Line scans across brazed joint of Fe-Co-Ni alloy & W brazed with OFE Copper & (b) Elemental distribution (Line Scans) of various elements with distance

EDS (Line scan) (Fig. 5(b)) results clearly revealed that the trend of distribution of copper and nickel followed the same pattern. This is also expected as Cu & Ni both forms an isomorphous system as per Hume Rothery rules [4]. Further, from Dot Map results, Fig. 8 (c) and (d), it seems that both Fe and Ni have diffused to very low extent. It is to be noted that as solid solubility of tungsten in OFE copper is extremely limited [8] that is also depicted from EDS (Dot Map and Line scan results, Figs. 5, 6 and 7). So, in the present case, it is expected that only surface wetting [9] was achieved to a very marginal extent. It may be inferred that Cu-Ni isomorphous system has been the main bonding support during this joining. Micro-hardness measurements were also done

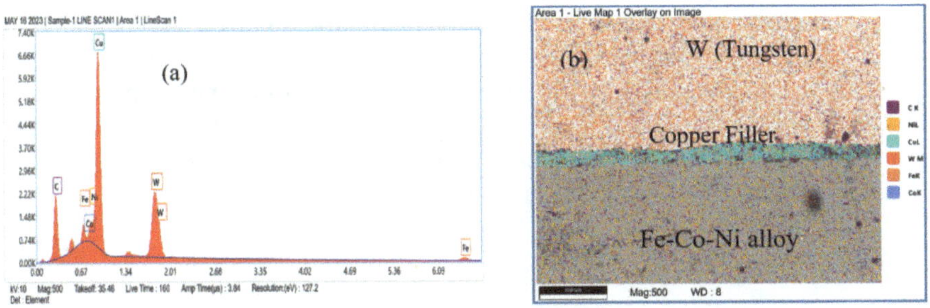

Fig. 6. (a) Elemental Peaks (area scan) obtained through EDS & (b) Dot Mapping of different elements shown collectively across brazed joint of Fe-Co-Ni alloy and tungsten

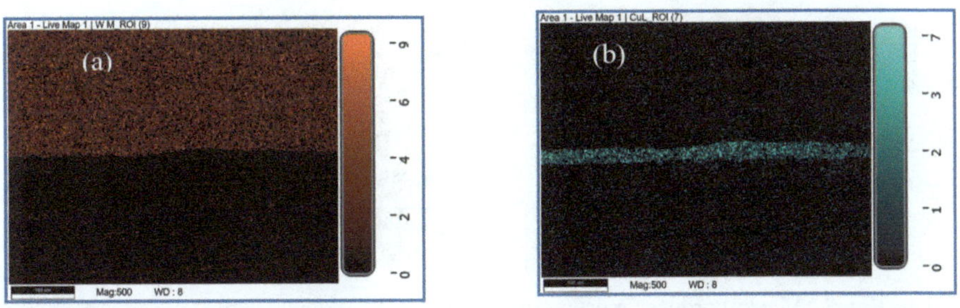

Fig.7. Dot maps of individual elements, viz, (a) tungsten and (b) copper in brazed joint of Fe-Co-Ni alloy and tungsten

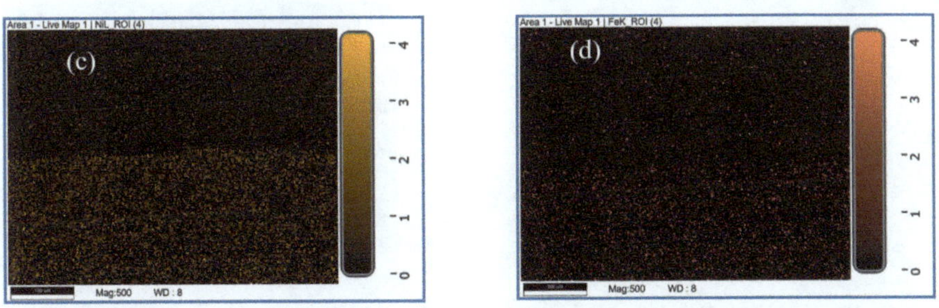

Fig. 8. Dot maps of individual elements, viz, (c) nickel and (d) iron in brazed joint of Fe-Co-Ni alloy and tungsten

across brazed region using Micro-hardness tester (Make: M/s Omni Tech, Pune) and the hardness was found to be lowest in filler region (~74 HV 0.01 kg) than parent materials (Fe-Co-Ni alloy: ~165 HV 0.01 kg & W: ~480 HV 0.01 kg) as shown in Fig. 9 with maximum deviations of about 4%, which indicates the soundness of interfacial joint using this process, as the filler alloy after brazing is still ductile enough and so will be able to accommodate the residual stresses easily [10–12].

During mechanical testing of Fe-Co-Ni alloy & tungsten brazed joint using UTM in compression mode, the maximum load obtained was 11450 N (Fig. 10) and corresponding joint strength obtained was 101 MPa. The mean value of joint strength was found to be 91.7 MPa, while the range of joint strength varied between 83 MPa to 101 MPa. The filler material layers (Fig. 11) were found on both parts of fractured samples that clearly indicated about the failure occurring at the joint.

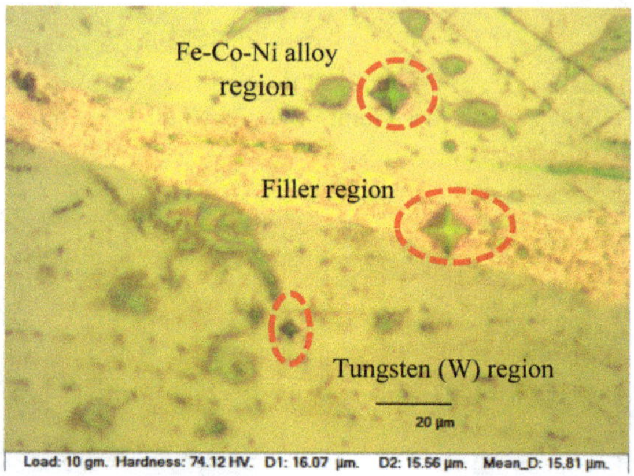

Fig. 9. Micro-hardness measurement (Vickers Scale) across brazed section

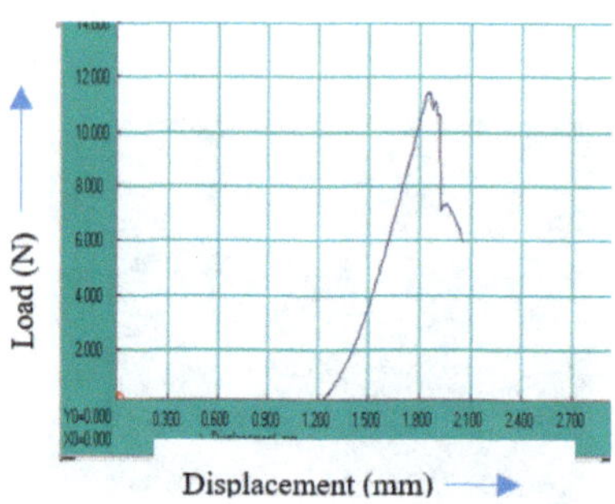

Fig. 10. Load V/s Displacement curve

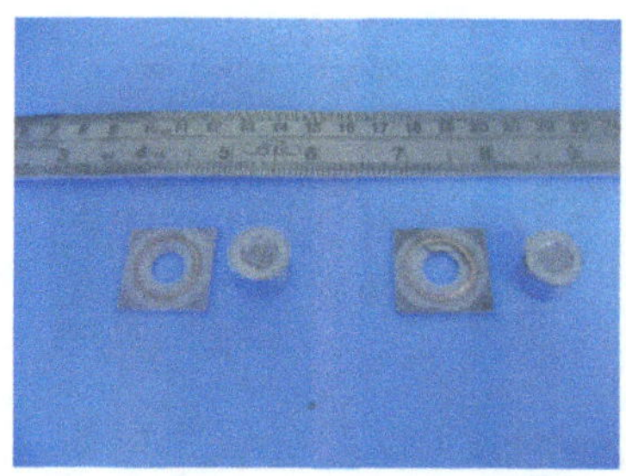

Fig. 11. Fractured samples after tensile test

4 Conclusions

Vacuum brazing of Tungsten and Fe-Co-Ni alloy using OFE Copper has been carried out successfully to form a micro-crack free, continuous and homogeneous joint. The microstructure revealed extremely limited diffusion of copper towards tungsten, while owing to mutual solubility of Cu & Ni in all proportions, Cu diffused towards Fe-Co-Ni alloy. Nickel, on account of Cu-Ni being an isomorphous system, has shown the same trend of distribution as that of copper. Further, the micro-hardness values clearly depicted that the joint quality is good as the filler region has least hardness signifying its ability to accommodate the residual stresses by undergoing plastic deformation. It is also to be noted that the brazed joint strength as obtained in this case was satisfactory, as this value is quite good enough for the application where this joint will be used in sub assembly component. Further, dead weights optimization for obtaining enhanced joint thickness and other properties will be studied and analyzed in future.

References

1. Recent advances in brazing fillers for joining of dissimilar materials. Metals **11**(7), 1037 (2021).https://doi.org/10.3390/met11071037
2. Yue, X., He, P., Feng, J.C., Zhang, J.H., Zhu, F.Q.: Microstructure and interfacial reactions of vacuum brazing titanium alloy to stainless steel using an AGCUTI filler metal. Mater. Charact. **59**(12), 1721–1727 (2008)
3. Brazing Fundamentals: http://www.asminternational.org, Chapter 2, 7–9
4. Schwartz, M.M.: Brazing (Second edition), 4–6, 11–13
5. Deng, N., Zhao, J., Yang, L., Zheng, Z.: Effects of brazing technology on hermeticity of alumina ceramic-metal joint used in nuclear power plants. Front. Mater. 7 (2020)
6. Xia, Y., Li, P., Hao, X., Dong, H.: Interfacial microstructure and mechanical property of TC4 titanium alloy/316L stainless steel joint brazed with Ti-Zr-Cu-Ni-V amorphous filler metal. J. Manufact. Processes, 35, 382–395 (2018)
7. Min, M., Mao, Y., Deng, Q., Wang, G., Wang, S.: Vacuum brazing of Mo to 316L stainless steel using BNi-2 paste and Cu inter-layer. Vacuum **175**(8), 109282 (2020)
8. Rao, M., Zhang, J., Ding, J., Luo, G., Shen, Q., Zhang, L.: Low-temperature diffusion bonding of W/Mo joints with a thin Cu interlayer. J. Mater. Process. Technol. **262**, 422–429 (2018)
9. Prado, J.de., Sánchez, M., Ureña, A.: Wettability study of a Cu-Ti alloy on tungsten and EUROFER substrates for brazing components of demo fusion reactor. Mater. Design **99**, 93–101 (2016)
10. Sene, F.F., Motta, C.C.: Synthesis and characterization of Ni-Mo filler brazing alloy for Mo-W joining for microwave tube technology. Mater. Res. **16**, 417–423 (2013)
11. Dong, H., Yang, Z., Wang, Z., Deng, D., Dong, C.: Vacuum brazing TC4 titanium alloy to 304 stainless steel with Cu-Ti-Ni-Zr-V amorphous alloy foil. J. Mater. Eng. Perform. **23**(10), 3770–3777 (2014)
12. Voiculescu, I., Geanta, V., Binchiciu, H., Iovanas, D., Stefanoiu, R.: Dissimilar brazed joints between steel and tungsten carbide. IOP Confe. Ser.: Mater. Sci. Eng. **209**, 012021 (2017)

Corrosion and High-temperature Oxidation

Water Parameters' Impact on the Corrosion Behaviour of High-Carbon Pearlitic Rail Steel

Kirtiratan Godbole[1](✉) and K. Mondal[2]

[1] Department of Applied Mechanics, Motilal Nehru National Institute of Technology, Allahabad, Prayagraj 211004, India
kgodbole@mnnit.ac.in

[2] Department of Materials Science and Engineering, Indian Institute of Technology Kanpur, Kanpur 208016, India

Abstract. Rail steels are high-carbon steels with a predominantly pearlitic microstructure that are commonly utilized in locomotive tyres and big forging dies. The study looks at how salinity, conductivity, total dissolved solids (TDS), and pH affect the corrosion of high-carbon pearlitic rail steel. The research includes pearlitic steels with coarse, fine, and very fine microstructures produced by furnace-cooled, air-cooled, and forced-cooled processes, as well as bainite, martensite, and tempered martensite produced by austempering, water quenching, and tempering of quenched steel, respectively. The research also looks at how water characteristics affect the composition, fractionation, and morphology of corrosion products. In general, salinity and conductivity rise as pH decreases over time. However, the increased TDS and protective index (α/γ^*) as measured by the weight fraction of stable α-FeOOH over the other unstable γ-FeOOH and β-FeOOH rust phases eventually induce enhanced corrosion resistance to the steels regardless of their microstructure and interlamellar spacing between ferrite (α-Fe) and cementite (Fe_3C). In all situations, air-cooled steel with fine pearlite outperforms furnace-annealed and forced-air-cooled steels. Furthermore, the bainitic microstructure steel is more corrosion resistant than the martensite and tempered martensite microstructure steels.

Keywords: Corrosion · Steel · Salinity · Conductivity · Martensite

1 Introduction

Steel degradation caused by corrosion is a critical issue in many industries, resulting in large financial losses. Corrosion-related issues are caused by a variety of factors, including adverse environmental conditions and insufficient maintenance. Corrosion is influenced by temperature, velocity, oxygen content, and the presence of living organisms in the electrolyte. Steel corrosion is a major problem for many industries, and it is impacted by a variety of factors such as microstructure, composition, and environment [1]. Because pearlite gives the best strength-ductility ratio, high-carbon steel with a totally pearlitic microstructure is commonly utilized as rail steel. Because pearlite has

A. Tiwari et al. (Eds.): iConFIRM 2023, SPPHY 308, pp. 127–139, 2024.
https://doi.org/10.1007/978-981-97-4557-9_14

the best strength-to-ductility ratio, rail steels manufactured of high-carbon steel with a completely pearlitic microstructure are commonly used. The bulk of rail steels are high-carbon steels with predominantly pearlitic microstructures, and they are frequently used in big forging dies, locomotive tyres, springs, and high-strength rope for bridge applications [2]. Heat treatment techniques can enhance the mechanical properties of high carbon steel by altering the pearlitic morphologies as well as the steel's whole microstructure without affecting its composition [3]. Rail steel maintenance and the accompanying costs of inspection are significant difficulties for modern railway systems. Because corroded rail lines must be replaced on a regular basis to avoid sudden rail collapse and traffic interruption, rail steel corrosion adds to the cost burden [4].

The fineness of the interlamellar spacing of pearlite, on the other hand, would have a significant influence on its corrosion capabilities. Pearlite forms a micro-galvanic cell due to the action of cementite and ferrite lamellae, which serve as cathode and anode, respectively [5, 6]. The ferrite microstructure in low-alloy carbon steels exhibits lower corrosion potential than the ferrite-bainite microstructure in a 3.5% NaCl solution [7]. In chloride-containing environments, bainitic steel outperforms martensitic steel [8]. Different heat treatment processes that change the composition of the same steel, such as normalizing, isothermal transformation, spheroidizing, oil quenching, and tempering, result in unique microstructures with altered phase size, shape, and distribution.

However, temperature, dissolved oxygen, electrolyte conductivity, and salinity all have a substantial influence on how a material performs in the submerged zone of saltwater [9, 10]. The electrolyte's total dissolved solids (TDS), the other determining parameter, is composed of organic and inorganic components in colloidal, particle, and ionic phases [11]. According to R. Starosta [12], as seawater salinity grows, so does austenitic steel's electro-chemical activity and corrosion susceptibility. In fact, nearly every investigation has shown that conductivity increases with salinity. As a result, steels would rust faster. According to K. Zakowski et al. [13], numerous water properties, such as conductivity, salinity, pH, oxygenation, total dissolved solids content, resistivity, and corrosion rate, are interconnected. The three distinct places are above the ocean surface, half immersed in seawater, and totally submerged in seawater. S. A. Al-Fozan et al. [9] investigated the effect of seawater level on the corrosion behaviour of different alloys in laboratory immersion experiments.

Research studying the combined effects of salinity, conductivity, total dissolved solids, and pH on the corrosion behaviour of diverse microstructures generated from a single composition of high-carbon rail steel in aerated 3.5% NaCl, on the other hand, would be exciting. Furthermore, it is interesting to investigate how water variables impact diverse steels, such as varied pearlitic morphologies and microstructures, such as bainitic, martensitic, and tempered martensitic steels formed from a same composition. Furthermore, rail lines are used in particularly hostile coastal settings. However, the morphologies of the various rust stages are critical for corrosion prevention. As a result, the research would help researchers better understand the corrosion properties of various microstructure steels. These six steels were created for the current study by heat-treating Indian rail steel. It should be emphasized that little research has been conducted on the influence of water properties on steel corrosion behaviour. The current study intends to correlate water parameters with the corrosion rates of the three high-carbon rail steel microstructures listed above utilizing an immersion test in a freely aerated 3.5% NaCl solution.

2 Experimental Procedure

The chemical composition of high-carbon Indian rail steel has been measured using Optical Emission Spectroscopy (Spectra Maxx) equipment. It is made up of the following components: C: 0.7, Si: 0.24, Mn: 1.12, P: 0.026, S: 0.021, Cr: 0.015, Ni: 0.012, As: 0.013, Nb: 0.013, Ta: 0.0725, and Fe (in wt.%) balance. The 30 × 30 × 3 mm^3 samples of high-carbon Indian rail steel was processed using electrical discharge machining (EDM). The microstructure of the high-carbon rail steel as received was completely pearlitic. Modifying the cooling rate produced a variety of pearlitic morphologies, including coarse, fine, and extremely fine structures. The sectioned high-carbon rail steel was subjected to three independent heat treatments in the second set of heat treatments to generate bainitic, martensitic, and tempered martensitic steels, respectively. Figure 1 shows a time-temperature-transformation curve of this high-carbon rail steel manufactured by H. Bhadeshia's programme [14]. It is used to produce various morphologies and microstructures of pearlitic steel.

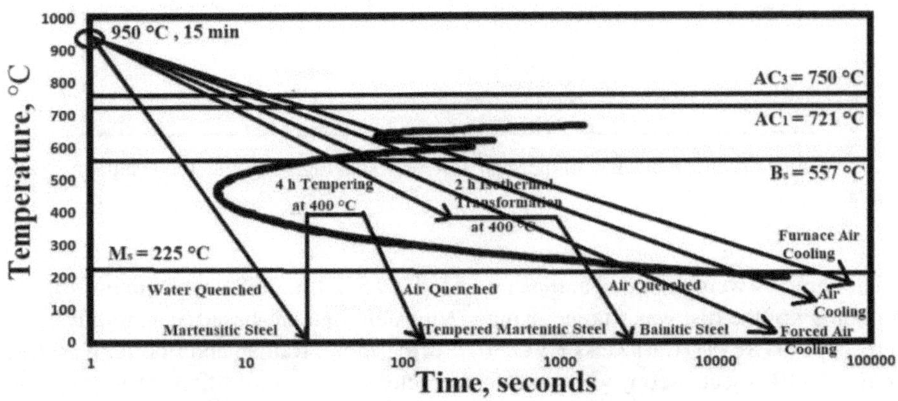

Fig. 1. Time-temperature-transformation diagram showing the schematic heat treatment cycles that generate distinct microstructures from high-carbon rail steel.

After being machined, the rail steel samples were austenitized for 15 min at 950 °C. They were then cooled to room temperature using one of three different methods: forced air cooling, air cooling, and furnace annealing (fast, medium, and slow). The sectioned rail steel samples were quenched at different cooling rates after being austenitized for 15 min at 950 °C to generate a range of microstructures. Water quenching was used to generate bainitic steel after two hours of austempering at 400 °C in a high-temperature salt bath containing 25% KCl + 20% NaCl and $BaCl_2$. Similarly, water quenching was utilized after austenitization to generate martensitic steel. The water-quenched samples were heated in a muffle furnace for four hours at 400 °C to make tempered martensitic steel.

Metallographic tests were performed on heat-treated steel samples. The steel samples were polished with 400–2500 grit silicon carbide paper before being cloth polished with 0.05 μm alumina powder dissolved in water. A 3% Nital solution was used to etch the mirror-polished samples in order to disclose the microstructures. The immersion corrosion investigation, depicted in Fig. 2, was carried out at room temperature using a freely aerated 3.5% NaCl solution for exposure periods of 5, 10, 15, 20, and 28 days. Prior to immersion testing, the exposed side of the heat-treated samples were polished using silicon carbide paper with a grain size of 2000. In preparation for the immersion test, the samples were polished and thoroughly cleaned with ethanol before being weighed. To coat and preserve the edges and unpolished surface, teflon tape and ordinary lacquer were utilized.

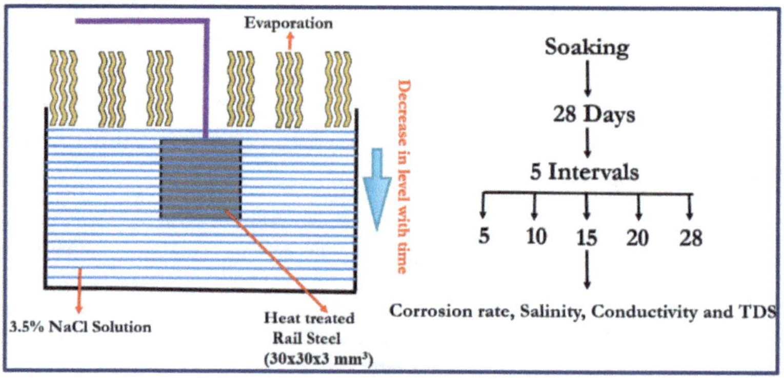

Fig. 2. An illustration of the lab immersion test and soaking schedule for evaluating water parameters.

The samples were obtained at the end of 5, 10, 15, 20, and 28 days of exposure. The morphology of the rust was studied using a Tungsten filament-based Scanning Electron Microscope (WSEM) (Carl Zeiss EVO 50). Furthermore, Raman and Fourier transform Infrared (FTIR) spectroscopy were used to characterize the composition of rust generated on the surface of various heat-treated samples after a defined exposure duration. Acton Spectra Pro SP-2500 Raman spectrometer (Princeton Instruments, STR Raman, TE-PMT detector) and Olympus optical microscope were used to produce Raman spectra. Using a Nd-YAG laser with a wavelength of 532 nm, the study was performed to analyze distinct stages of the corroded components. The FTIR spectra may be used to do both qualitative and quantitative investigation to determine the relative proportions of distinct rust stages. The protective ability index (α/γ^*) of a corroded surface made of hydrated iron was evaluated using the FTIR spectrum. The rust coating on the samples was removed after characterization by immersing them in Clarke's solution (20 g of Sb_2O_3 and 50 g of $SnCl_2$ in 1000 ml of HCl), followed by five minutes of ultrasonic cleaning with distilled water and acetone. The final weight was then calculated in order to assess the corrosion rate in accordance with (ASTM-G1-03) [15]. The water parameters of the solution containing 3.5% NaCl used in the immersion test were assessed after each exposure period of 5, 10, 15, 20, and 28 days. A Hanna EDGE Salinity Metre was used to determine total dissolved solids (TDS) (g/L), pH, conductivity (mS/cm), and salinity (g/L).

3 Results and Discussion

Figure 3(a–f) shows SEM micrographs of the heat-treated materials. The micrographs for the air-cooled, forced-air-cooled, and furnace-annealed steel samples are shown in Figs. 3(a–c). These micrographs show a completely developed lamellar pearlitic microstructure. The characteristic non-lamellar feather-like microstructure morphology of bainitic rocks is seen in Fig. 3(d). Figure 3(e) shows a martensitic microstructure with needle-like long, non-oriented morphological characteristics. Figure 3(f) depicts the microstructure of bushy-tempered martensitic steel. Carbides begin to precipitate in the fine-grained ferritic matrix around the plate boundaries of the needle-shaped ferrite morphology.

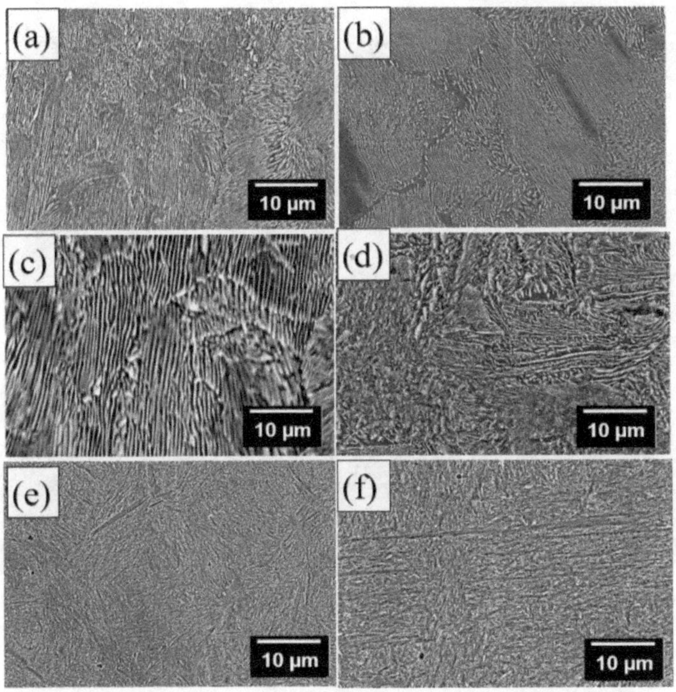

Fig. 3. SEM micrographs of the different morphologies of high-carbon pearlite steels: (a) furnace-annealed, (b) air-cooled, (c) forced-air-cooled, and different microstructures of high-carbon pearlite steels: (d) Bainite, (e) Martensite and, (f) Tempered Martensite.

Figure 4 shows SEM micrographs of the corroded surfaces of all six pearlitic steel types without etching during the initial five-day immersion test exposure period. Rust products with different morphologies, such as Goethite (α-FeOOH), Lepidocrocite (γ-FeOOH), Akaganeite (β-FeOOH), Feroxyhyte (δ-FeOOH), and Magnetite (Fe_2O_3), are often found on steels under immersion conditions [16]. Figure 4(a–f) shows SEM micrographs of the corroded surfaces of air-cooled, forced-air-cooled, furnace-annealed, bainite, martensite, and tempered Martensite samples made from heat-treated pearlitic steel samples for early-stage immersion during a 5-day period. Figure 4(a & c) reveals fragmented rust morphology all over the surfaces of air-cooled and forced-air-cooled tanks at the early stage of immersion (5 days). Furthermore, the micrograph Fig. 4(b & e) shows a cotton ball-type rust outcome, which may be characterized as goethite (α-FeOOH) [17–19]. Figure 4(d & f) depicts the cloud-like and needle-like shape of the α-FeOOH phase in bainite and tempered martensite, respectively. The corrosion products on the rusted surface are not constant throughout the first five days of the immersion test.

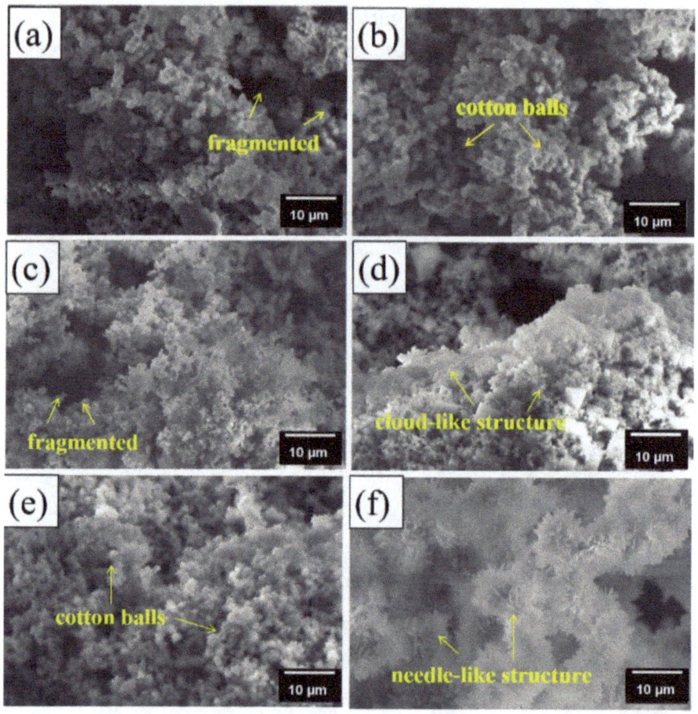

Fig. 4. SEM micrographs of rust product after early stage of immersion (5 days) on (a) furnace-annealed, (b) air-cooled, (c) forced-air-cooled, (d) Bainite, (e) Martensite and, (f) Tempered Martensite.

Figure 5(a–d) depicts SEM micrographs of corroded surfaces with varying morphologies and microstructures acquired after a 28-day final exposure period. Figure 5(d) shows that the number of these needle-like formations grew during the course of the 28-day immersion. The various crystalline aggregate sizes and shapes seen in the observed rust morphologies illustrate how immersion circumstances impact corrosion. Figure 5(b) shows the surface consisting of thick and dense layers of rust components after a 28-day immersion test, with the α-FeOOH phase's unique cotton ball and whiskers form apparent (Fig. 5(c)). However, after 28 days of immersion, the whole martensite and tempered martensite steel sample formed an equal coating of whiskers (Fig. 5(e&f)). But after 28 days in the water (Fig. 5(a–f)), the rusts are uniformly distributed over the surface. It has a striking resemblance to goethite's needle-like structures and characteristic whiskers (α-FeOOH) [20].

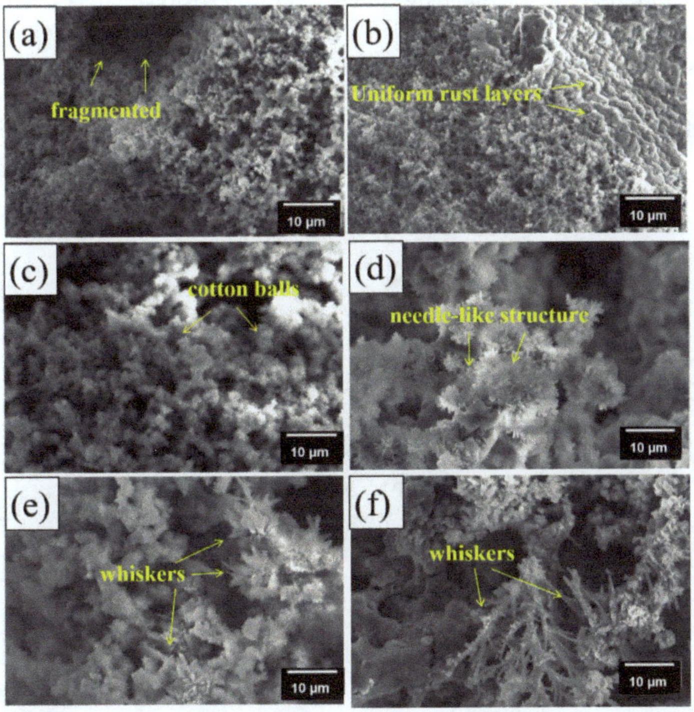

Fig. 5. SEM micrographs of rust product after the final stage of immersion (28 days) on (a) furnace-annealed, (b) air-cooled, (c) forced-air-cooled, (d) Bainite, (e) Martensite and, (f) Tempered Martensite.

The surfaces of the different morphologies, namely furnace-annealed, air-cooled, forced-air-cooled, and the different microstructures, namely bainitic, martensitic, and tempered martensitic steels, produced rust components after 5, 10, 15, 20, and 28 days of exposure, as shown by their Raman spectra in Fig. 6(a–f). Raman spectroscopy of the constituents of rust confirms the primary corrosion product of γ-FeOOH (Lepidocrocite) (peaks at 254 and 1310 cm^{-1}) and other peaks such as α-FeOOH (Goethite) (peak at 1085 cm^{-1}) and δ-FeOOH (Feroxyhyte) (peaks at 380 and 670 cm^{-1}) and β-FeOOH (Akaganeite) (peak at 538 cm^{-1}). The FTIR spectra given in Fig. 6(g–m) demonstrate the occurrence of phases such as $Fe(OH)_2$, β-FeOOH, α-FeOOH, and γ-FeOOH.

Peaks appear at 482, 674, 870, and 1020 cm^{-1} on the surface of the rust component in all three high-carbon rail steel microstructures. These larger and deeper dips are extremely close to their typical values [21]. Peaks with higher wave numbers arise as a result of O-H stretching [22]. FTIR and Raman spectroscopy investigations give strong mutual confirmation. The quantity of rust components that form on corroded surfaces varies depending on the microstructure immersed in freely aerated 3.5% NaCl for 28 days. The protective ability index, computed from the FTIR spectra of the rust components using the stable phase to dissolving rust phase ratio (α/γ*), quantifies how effective the rust layer is [23, 24]. The α/γ* ratio, which relates to rust's protective ability, is an important metric that determines steel's corrosion tendency. This ratio may be calculated using the FTIR spectra of rust elements by dividing the weight fraction of stable phases by the weight fraction of dissolving rust phases. The protective capacity of rust is determined by the quantitative ratio of alpha/gamma* (goethite/(lepidocrocite + akaganeite)), or the mass of crystalline α-FeOOH to the total mass of γ-FeOOH and β-FeOOH in the rust layer [25].

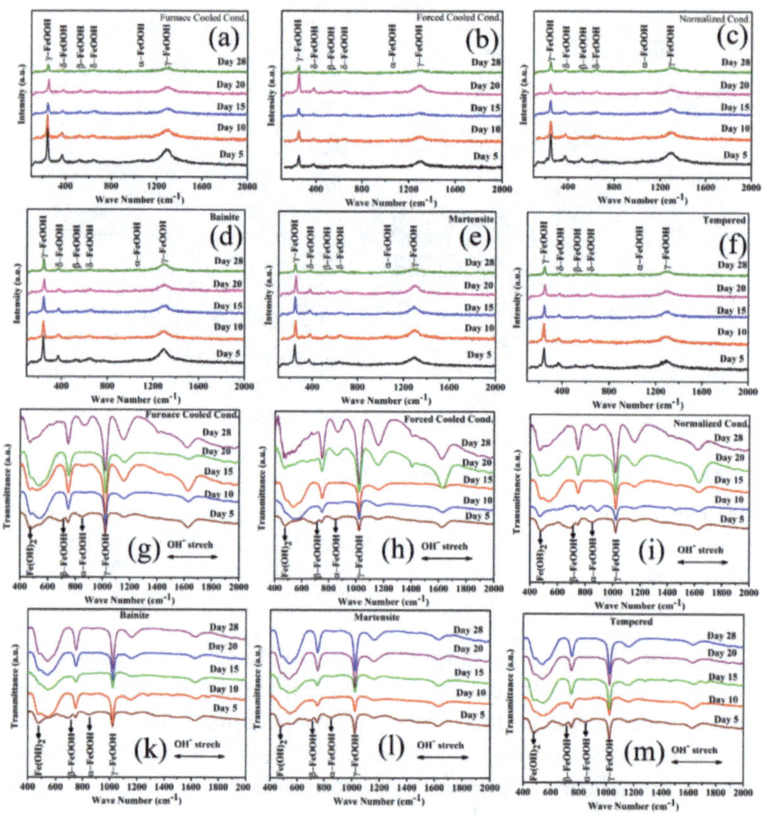

Fig. 6. Raman spectrum of the rust constituent formed after immersion on (a) furnace-annealed, (b) air-cooled, (c) forced-air-cooled, (d) Bainite, (e) Martensite, (f) Tempered Martensite, and FTIR spectrum of the rust constituent formed after immersion on (g) furnace-annealed, (h) air-cooled, (i) forced-air-cooled, (k) Bainite, (l) Martensite, (m) Tempered Martensite.

Figure 7(a–c) visually depicts salinity (g/L), conductivity (mS/cm), TDS (g/L), pH, corrosion rate (mpy), and α/γ* ratio with exposure time. The oscillations demonstrate a well-known general rise in water parameters, with the exception of pH, which declines over the course of the exposure days. Water evaporation increases salinity, conductivity, and TDS as exposure time increases [26]. The hydrolysis of the dissolving iron ion, on the other hand, results in the formation of HCl acid. In Fig. 7(b & c), it has been demonstrated that an increase in the ionic conductivity, salinity level, and TDS of the aqueous NaCl solution corresponds with the rust protecting potential and the rate of corrosion of distinct pearlitic morphologies and microstructure steel variants. Steel corrosion rate increases proportionally with salt in natural water; beyond 3% salinity, oxygen solubility reduces, which tends to reduce corrosion rate in seawater. Low oxygen concentrations come from high salinity values in saltwater caused by evaporation, which is why oxygen solubility is critical in limiting corrosion rate. All of the research shows that increasing the salinity and conductivity of steel increases the corrosion rate [27, 28].

The current study, on the other hand, discovered an inverse relationship between salinity and conductivity (two water parameters) and corrosion rate. Furthermore, as seen in Fig. 7, extended exposure intervals lower corrosion rate while increasing TDS and α/γ^* ratio. The major reason of the TDS increase is an increase in iron hydroxide/oxyhydroxide concentrations. However, when Fe dissolves in the chloride medium, the hydrolysis of $FeCl_2$ or $FeCl_3$ produces iron hydroxide/oxyhydroxide and HCl (strong acid), which causes the pH to decrease. Due to the pH dropping with exposure time, γ-FeOOH has transformed into a non-reactive α-FeOOH phase. According to studies, in somewhat acidic settings, the predominant phase of steel rust is lepidocrocite, which eventually transforms into the more stable hydroxide goethite [29]. A rise in TDS and a reduction in pH result in the formation of a stable geothite phase (α-FeOOH). This evolution is consistent with FTIR and Raman spectroscopy (Fig. 6), as well as the rust morphologies described in Figs. 4 and 5. As a result, the mass ratio of the reactive (γ-FeOOH) and non-reactive (α-FeOOH) phases may be used to calculate the TDS's capacity to protect.

Furthermore, the prerequisite is spontaneous evaporation without the addition of a new solution or solvent. The water evaporation phenomena increases conductivity and salinity with exposure time. Surprisingly, the TDS of the electrolyte rises with exposure time. Because distilled water was used to prepare the NaCl solution as well, no NaCl was added during the experiment. The major cause of the TDS rise might be the increasing iron-based salt, which is mostly hydroxide or oxyhydroxide of iron. Therefore, when the TDS increases, the interactions between rust and dissolved solids attempt to progress towards the production of more compact rusts. TDS and the rust's protectivity index (α/γ^* ratio) are the most important factors in regulating the corrosion rate with exposure duration. The hydrolysis process produces iron hydroxide and a strong HCl acid, which leads in a pH drop. As a consequence, after 28 days of immersion, a rise in salinity, conductivity, and TDS, along with a reduction in pH, has resulted in lower corrosion susceptibility of pearlitic steels, regardless of pearlitic morphology.

During immersion, iron oxidation occurs through the cathodic reduction reaction of γ-FeOOH (Lepidocrocite). This is supported by the decreased intensity of the γ-FeOOH peak in Raman spectra (Fig. 6(a–f)) at 254 and 1310 cm^{-1} with exposure days in all variants. The FTIR spectra (Fig. 6(g–m)) show a reduction in peak intensity at 1020 cm^{-1} with exposure days for all pearlitic steel types. It appears that the PAI of the rust layer

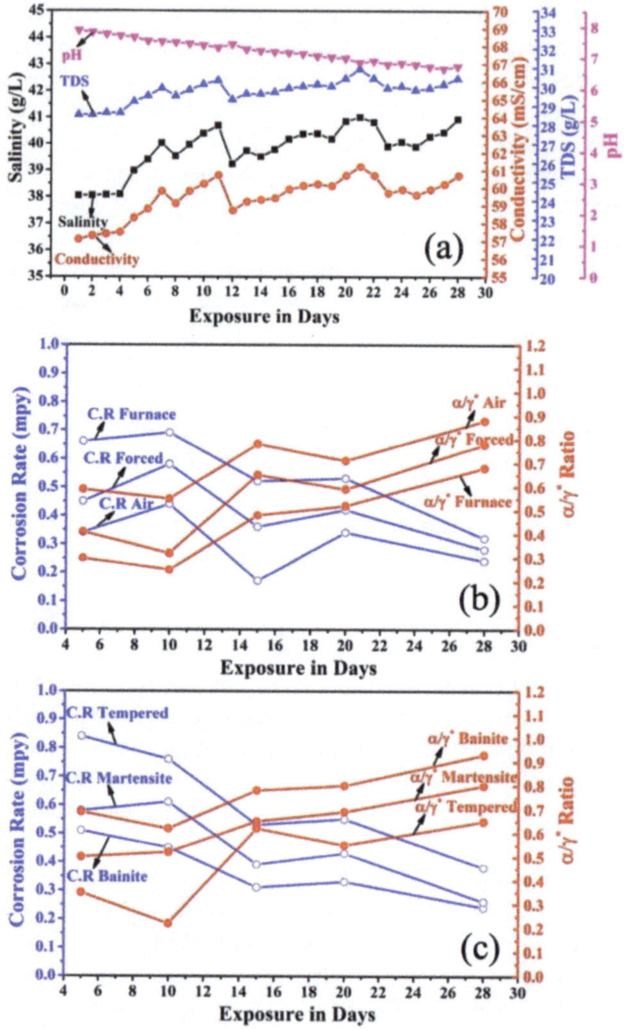

Fig. 7. Graphical representation of (a) salinity (g/L), conductivity (mS/cm), TDS (g/L), and pH, (b) corrosion rate, and α/γ* of different morphologies and (c) corrosion rate, and α/γ* of different microstructures of high-carbon rail steel with exposure in days immersed in 3.5% NaCl.

rises with exposure period in all variations. SEM micrographs revealed the cotton ball and whisker shape, which is commonly associated with the stable α-FeOOH (Goethite) phase (Figs. 4 and 5), regardless of pearlitic variations and different microstructures. In general, α-FeOOH and $Fe(OH)_2$ are the rust phases that contribute to the protective layer and may be recognised using FTIR spectra, as illustrated in Fig. 6. Similar observations about a rise in the α/γ* ratio and the stability of the protective layer of α-FeOOH over γ-FeOOH have been reported in the literature [30, 31].

As a result, a decrease in pH and an increase in TDS and the α/γ^* ratio induce the corrosion rate to trend lower, even when conductivity and salinity increase with exposure time, owing mostly to natural evaporation. Six steels are chosen specifically to illustrate that this element of corrosion rate variation is not unique to any one steel. A rise in TDS and the α/γ^* ratio, an increase in conductivity and salinity due to natural evaporation, and a decrease in pH are all variables that contribute to this volatility. In addition, the lack of fresh solvent or solution addition, as well as spontaneous evaporation, are required. Figure 7(b & c) depicts the significant association between the α/γ^* ratio and the corrosion rate of different pearlitic steel morphologies and microstructures. The correlation between corrosion rate and α/γ^* ratio is commonly accepted. Finally, it has been established for the first time that the TDS and rust protectivity (α/γ^* ratio) of the chloride electrolyte may offset the increase in salinity and conductivity, as well as the decrease in pH, and slow the rate at which steel corrodes independent of its microstructure. Furnace-annealed steel has the lowest protective index (α/γ^*), whereas air-cooled steel has the highest. Forced-air-cooled steel has an intermediate protective index (α/γ^*). The corrosion resistance of pearlite steels is determined by interlamellar spacing. The protective index value (α/γ^*) increases as the interlamellar spacing decreases from furnace-annealed to air-cooled steel, except for forced-air-cooled steel, which has begun to decrease throughout the exposure duration. This might be owing to a greater degree of breaking of the alternative cementite lamellae, resulting in an increase in the physical area of the galvanic cell as a result of micro-galvanic cell shorting out. This breaks the surface's protectiveness as compared to the air-cooled. Forced-air-cooled and furnace-annealed steels are less corrosion resistant than air-cooled steel with fine pearlite under all conditions.

While comparing the corrosion rates of the different microstructure steels, Fig. 7(c) shows that the bainitic steel's α/γ^* ratio is always higher than that of the martensite and tempered martensite throughout the immersion time. In any carbon steel microstructure, ferrite and cementite operate as anodes and cathodes, respectively, resulting in the development of a micro-galvanic cell. However, the carbide distribution in the microstructure causes variations in the galvanic couple length scales. Bainitic steels contain carbides of smaller length scale (Fig. 3d), resulting in a decreased corrosion rate. In contrast, both martensitic and tempered martensitic steel (Fig. 3e, f) have a larger contact area of ferrite and cementite phases due to a linked network of cementite, resulting in inferior corrosion resistance. Therefore, bainitic steels have higher corrosion resistance than martensite and tempered martensite steels. Specifically, the current study uncovers fresh discoveries showing a rise in TDS combined with a reduction in pH outweighs an increase in conductivity and salinity, imparting corrosion resistance to all three pearlitic steel varieties.

4 Conclusions

This study investigates the salinity, conductivity, TDS, pH, and corrosion rates of diverse morphologies and microstructures produced from a single composition in a freely aerated 3.5% NaCl solution. It discovers that corrosion rates have dropped as salinity, conductivity, and TDS have increased, as well as the pH of the electrolyte has fallen to a more acidic state. The only factors influencing the reduction in steel corrosion rate are an increase in TDS and rust protectivity (/* ratio), which is independent of morphologies and microstructure, and occurs even as the electrolyte becomes acidic and salinity and conductivity rise with exposure. Corrosion resistance is lower in forced air cooled and furnace-annealed steels than in air-cooled steel with fine pearlite. Furthermore, the bainitic microstructured steel outperforms martensite and tempered martensite microstructures in terms of corrosion resistance.

References

1. Panda, B., Balasubramaniam, R., Dwivedi, G.: On the corrosion behavior of novel high carbon rail steels in simulated cyclic wet-dry salt fog conditions. Corros. Sci. **50**, 1684–1692 (2008)
2. Rault, V., Vignal, V., Krawiec, H., Tadjoa, O.: Corrosion behavior of heavily deformed pearlitic and brass-coated pearlitic steels in sodium chloride solutions. Corros. Sci. **86**, 275–284 (2014)
3. Al-rubaiey, S.I., Anoon, E.A., Hanoon, M.M.: The influence of microstructure on the corrosion rate of carbon steels. Eng. Tech. J. **31**, 1–12 (2013)
4. Hlavaty, I., Sigmund, M., Krejci, L., Mohyla, P.: The bainitic steels for rail applications. Mater. Eng. **16**(4), 44–50 (2009)
5. Katiyar, P.K., Misra, S., Mondal, K.: Effect of different cooling rates on the corrosion behavior of high-carbon pearlitic steel. J. Mater. Eng. Perform. **27**(4), 1753 (2018)
6. Katiyar, P.K., Misra, S., Mondal, K.: Comparative corrosion behavior of five microstructures (pearlite, bainite, spheroidized, martensite, and tempered martensite) made from a high carbon steel. Metall. Mater. Trans. A **50**, 1489–1501 (2019)
7. Qu, S., Pang, X., Wang, Y., Gao, K.: Corrosion behavior of each phase in low carbon microalloyed ferrite–bainite dual-phase steel: experiments and modeling. Corros. Sci. **75**, 67 (2013)
8. Kazum, O., Bobby Kannan, M.B., Beladi, H., Timokhina, I.B., Hodgson, P.D., Khoddam, S.: Aqueous corrosion performance of nanostructured Bainitic steel. Mater. Des. **54**, 67–71 (2014)
9. Al-Fozan, S.A., Malik, A.U.: Effect of seawater level on corrosion behavior of different alloys. Desalination **228**, 61–67 (2008)
10. Atashin, S., Pakshir, M., Yazdani, A.: Comparative studying of marine parameters' effect, via qualitative method. T. App. Sci. Res. **5**(2), 120–128 (2010)
11. Boerlage, S.F.E.: Measuring salinity and TDS of seawater and brine for process and environmental monitoring-which one, when? Desalin. Water Treat. **42**(1–3), 222–230 (2012)
12. Starosta, R.: Influence of seawater salinity on corrosion of austenitic steel. J. Kones Power. Trans. **26**, 3 (2019)
13. Zakowski, K., Narozny, M., Szocinski, M., Darowicki, K.: Influence of water salinity on corrosion risk—the case of the southern Baltic Sea coast. Environ. Monit. Assess. **186**, 4871–4879 (2014)
14. Bhadeshia, H.: Materials Algorithm Project Program Library, Phase Transformation Group, Department of Materials Science and Metallurgy, University of Cambridge, Cambridge, U K. https://www.phasetrans.msm.cam.ac.uk/map/steel/programs/mucg46-b.html
15. Moon, A.P., Sangal, S., Layek, S., Giribaskar, S., Mondal, K.: Corrosion behavior of high-strength bainitic rail steels. Metall. Mater. Trans. A **46A**, 1500–1518 (2015)
16. Wu, H., Lei, H., Chen, Y.F., Qiao, J.: Comparison on corrosion behaviour and mechanical properties of structural steel exposed between urban industrial atmosphere and laboratory simulated environment. Constr. Build. Mater. **211**, 228–243 (2019)

17. Surnam, B.Y.R., Chui, C.W., Xiao, H., Liang, H.: Investigating atmospheric corrosion behavior of carbon steel in coastal regions of Mauritius using Raman Spectroscopy. Rev. Mater. **21**, 151–168 (2016)
18. Raman, A., Nasrazadani, S., Sharma, L.: Morphology of rust phases formed on weathering steels in various laboratory corrosion tests. Metallography **22**, 79–96 (1989)
19. Antunes, R.A., Costa, I., de Faria, D.L.A.: Characterization of corrosion products formed on steels in the first months of atmospheric exposure. Mater. Res. **6**, 403 (2003)
20. Dillmann, Ph., Mazaudier, F., Hoerle, S.: Advances in understanding atmospheric corrosion of iron: I. Rust characterisation of ancient ferrous artefacts exposed to indoor atmospheric corrosion. Corros. Sci. **46**(6), 1401 (2004)
21. Balasubramaniam, R., Ramesh Kumar, A.V., Dillmann, P.: Characterization of rust on ancient Indian iron. Curr. Sci. **85**, 1546 (2003)
22. Balasubramaniam, R., Ramesh Kumar, A.V.: Characterization of Delhi iron pillar rust by X-ray diffraction, Fourier transform infrared spectroscopy and Mossbauer spectroscopy. Corros. Sci. **42**, 2085–2101 (2000)
23. Fonna, S., Ibrahim, I.B.M., Gunawarman, Huzni, S., Ikhsan, M., Thalib, S.: Investigation of corrosion products formed on the surface of carbon steel exposed in Banda Aceh's atmosphere. Heliyon **7**, e06608 (2017)
24. Marcotte, T.D., Hansson, C.M.: Corrosion products that form on steel within cement paste. Mater. Struct. **40**(3), 325–340 (2007)
25. Waseda, Y., Suzuki, S. (eds.): Characterization of Corrosion Products on Steel Surfaces. Springer, Berlin (2006)
26. Da-Allada, C.Y., Alory, G., du Penhoat, Y., Kestenare, E., Durand, F., Hounkonnou, N.M.: Seasonal mixed-layer salinity balance in the tropical Atlantic ocean: mean state and seasonal cycle. J. Geo. Res. Oceans **118**, 332–345 (2013)
27. Toloei, A., Atashin, S., Pakshir, M.: Corrosion rate of carbon steel under synergistic effect of seawater parameters including pH, temperature, and salinity in turbulent condition. Corros. Rev. **31**(3–6), 135–144 (2013)
28. Sundjono, Priyaotomo, G., Nuraini, L., Prifiharni, S.: Corrosion behaviour of mild steel in seawater from Northern Coast of Java and Southern Coast of Bali, Indonesia. J. Eng. Technol. Sci. **49**(6), 770–784 (2017)
29. Stratmann, M., Bohnenkamp, K., Engell, H.J.: An electrochemical study of phase-transitions in rust layers. Corros. Sci. **1983**(23), 969–985 (1983)
30. Choudhary, S., Garg, A., Mondal, K.: Relation between open circuit potential and polarization resistance with rust and corrosion monitoring of mild steel. J. Mater. Eng. Perform. **25**, 2969–2976 (2016)
31. Yamashita, M., Nagano, H., Misawa, T., Townsend, H.E.: Structure of protective rust layers formed on weathering steels by long-term exposure in the industrial atmospheres of Japan and North America. ISIJ Int. **38**, 285–290 (1998)

Molecular Dynamics Simulations of Diffusivity in an Atomic Scale Diffusion Multiple

Jhalak(✉), Dishant Beniwal, and Pratik K. Ray

Indian Institute of Technology Ropar, Rupnagar, Punjab 140001, India
jhalak.20mmz0010@iitrpr.ac.in

Abstract. Accurate measurement of diffusivity is critical for several diffusion-controlled materials phenomena ranging from creep to oxidation. Experimental measurements of diffusion coefficients are expensive and time-intensive, especially while using diffusion couples. To an extent, the measurements are expediated using diffusion multiples. Ab-initio methods, within the framework of the Nudged-Elastic Band method as well as Molecular Dynamics (MD) can be employed for modeling diffusion. However, the traditional diffusion couple approach makes this time-consuming as well. Here, we adopt classical MD for creating an experimental analogue of diffusion multiples and demonstrate the concept using a ternary diffusion multiple consisting of Ti, Ni and V and assess the interdiffusion coefficients across Ti-Ni, Ti-V and Ni-V interfaces.

Keyword: Diffusivity; Diffusion multiple; Boltzmann-Matano

1 Introduction

Diffusion plays an important role in various materials phenomena such as microstructural changes and materials processing [1], creep, oxidation [2], etc. [3–7]. The diffusive phenomena are usually well-described by Fick's laws. Therefore, the diffusion coefficients are often extracted using Fick's second law or by using a Boltzmann-Matano (B-M) analysis (usually employed for diffusion couples). In this article, we have computationally simulated a ternary diffusion multiple where three B-M interfaces are present simultaneously from where the diffusion coefficients have been calculated.

Ficks' laws of diffusion are used to describe diffusion of species in solids [4, 8]. The first law relates the diffusion flux to the concentration gradient while the second law describes the temporal evolution of concentration and is often leveraged for estimating diffusivities using computational methods [9]. According to Fick's second law:

$$\frac{\partial C}{\partial t} = D\nabla^2 C$$

where, C is concentration, t is time of diffusion, D is diffusion coefficient which is given by the expression:

A. Tiwari et al. (Eds.): iConFIRM 2023, SPPHY 308, pp. 140–147, 2024.
https://doi.org/10.1007/978-981-97-4557-9_15

$$D = D_0 \exp\left(-\frac{Q}{RT}\right)$$

For diffusion of species, concentration profiles can be obtained by applying boundary conditions to the expression for Fick's second law. These concentration profiles are fitted with the experimentally obtained profiles to calculate diffusion coefficient or diffusivity. However, the accuracy of determination of diffusion coefficient depends on the type of solution to be used [10–12]. If D varies with the composition, a Boltzmann-Matano (B-M) solution is often employed.

Experimentally, concentration profiles are obtained using Electron Probe Microanalysis (EPMA). In the present work, concentration profiles were obtained using atomistic simulations in diffusion multiples. A diffusion multiple is a group of three or more distinct metal blocks that are in close contact with each other [13–15]. The geometry of diffusion multiples is created such that the maximum number of B-M interfaces, which allows for the measurement of multiple diffusivities from a single diffusion multiple. As the diffusion multiple is heated to high temperatures, interdiffusion occurs across multiple interfaces. Experimental assessments of diffusivity through diffusion multiples provide a high through-put approach for diffusivity measurements, but has its own unique challenges, primarily stemming from relatively high cost and time requirements. Additionally, it is difficult to extract diffusion coefficients for metastable materials and in-situ profiles at high temperatures. For instance, Zhu et al. [12] extracted concentration profiles experimentally using EPMA after hot isostatic pressing at different temperatures followed by quenching. Diffusion coefficients were extracted following high temperature annealing and quenching. However, the diffusivity data obtained was considered inaccurate in the vicinity of bcc Ti transforming to hcp (882 °C).

The aim of this work is to demonstrate a proof of concept whereby we use classical MD simulations to model a ternary atomic-scale diffusion multiple. Use of this technique enables to extract concentration profiles of heated system at any timestep. In addition to demonstrating the utility and efficacy of the approach, we also highlight the potential challenges that need to be overcome, primarily arising from modeling the geometry while allowing for sample drift.

2 Methodology

A Ni-Ti-V diffusion multiple was created by combining individual slabs of Ni, Ti and V using LAMMPS [16] and ATOMSK [17]. The dimensions of Ni, Ti and V slabs were 57.57 x 52.86 x 90.9 Å^3, 59.02 x 98.31 x 90.9 Å^3 and 57.57 x 45.45 x 90.9 Å^3 respectively resulting in the overall simulation box size of 115.5 x 98.31 x 90.9 Å^3 containing 73,566 atoms. Figure 1 shows the structure of Ni-Ti-V multiple that was used for the simulations. A modified embedded-atom method (MEAM) potential, developed by Maisel et al. [18], was used for the interdiffusion studies in ternary Ti-Ni-V system. Maisel et al. [18] validated the potential using the experimental thermal properties and melting temperatures of Ni, Ti and V elements and also studied the transformations in Ni-Ti-V using this potential. It is, therefore, expected to describe the system at the target temperature of 1000 °C. Vacancies were introduced by removing atoms at random positions in the individual element slabs while assuming Arrhenius relationship between the equilibrium number of vacancies and temperature. At each temperature, the structure was heated for 30 ps (using a step size of 0.001 ps) and the atomic positions estimated after fixed time steps – every 100 steps from 0–1,000 steps, every 1,000 steps from 1,000–10,000 time steps and every 10,000 time steps thereafter. The interdiffusion was studied by examining the concentration profiles across all three interfaces – Ti-Ni (Ifc_1), Ti-V (Ifc_2), and Ni-V (Ifc_3). Periodic and fixed boundary conditions were used along Z-axis and (X, Y)-axes respectively.

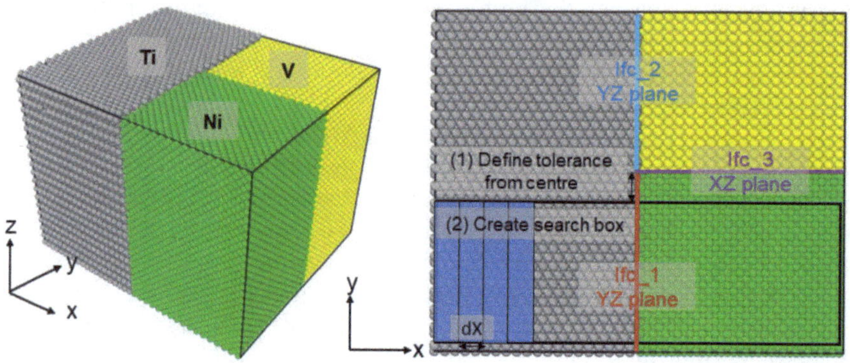

Fig. 1. Diffusion multiple

3 Results and Discussion

As the diffusion multiple is heated at a fixed temperature, atomic position snapshots were saved at regular time steps. Each of these snapshots were then analyzed to obtain the concentration profile across all interfaces. Figure 1 shows the visualization of XY plane of the diffusion multiple (at $t = 0$) as observed from Z-direction. The three interfaces that can be seen here are: i) Ifc_1 (Ti-Ni interface, YZ plane), ii) Ifc_2 (Ti-V interface, YZ plane) and iii) Ifc_3 (Ni-V interface, XZ plane). We will take the example of Ifc_1 here to describe the methodology used for obtaining concentration profiles and diffusivity values across any interface. First, we use a tolerance of 10Å from the triple junction and create a three-dimensional search box that spans the entire width across the Ifc_1 interface, as shown in Fig. 1. The 'width' represents the dimension normal to the interface plane (which is X-direction for Ifc_1). Then, we discretize the width of the search box using a step size of dx (= 2Å in this work) and create slices within the search box wherein each slice is parallel to YZ plane and has a width of 2Å. Within each slice volume, we count the number of atoms of each element as well as the total atoms that are present to calculate the elemental concentrations. Thus, by analyzing all slices, we obtain concentration values across an interface as a function of distance from interface, as shown in Fig. 2.

The next step is to fit the concentration values to a continuous concentration profile which can be then analyzed to obtain interdiffusion coefficients using the B-M analysis. Since the concentration profile can be non-symmetric about the Matano interface, we used a skew-cumulative distribution function (CDF) [19] to model the concentration profile curves across an interface. A skew-CDF is a combination of normal cumulative distribution function, φ and Owen's T function [20], T. Mathematically, the concentration profile $C(x)$ as a function of distance from interface (x) was modelled as:

$$C(x) = \varphi(X) - 2T(X, \alpha)$$

$$X = \frac{x - \xi}{\omega}$$

$$\varphi(X) = \frac{1}{2}\left(1 + \text{erf}\left(\frac{X}{\sqrt{2}}\right)\right)$$

$$T(X, \alpha) = \frac{1}{2\pi}\int_0^{\alpha} \frac{\exp\left(-0.5X^2\left(1 + h^2\right)\right)}{1 + h^2} dh$$

where, $\xi \rightarrow$location, $\omega \rightarrow$scale, $\alpha \rightarrow$shape factors of skew distribution and $x \rightarrow$distance from interface

The concentration profiles were fitted to the skew-CDF using the modified Powell algorithm (implemented in Python using scipy) which optimizes the ξ, ω and α parameters to minimize the error between the fit profile and calculated concentration values.

B-M interface analysis was then implemented on the continuous skew-CDF profiles at each interface to calculate the interdiffusion coefficient values. The first step in B-M analysis was to calculate the location of Matano plane which is such that the area under the concentration profile on one side of the Matano plane is equal to the area above the concentration profile on the other side of the Matano plane, as shown in Fig. 3a. Mathematically, the location of Matano plane was identified by solving the below equation for value of x_M:

$$\int_{x_L}^{x_M} C(x)dx = \int_{x_M}^{x_R} C(x)dx$$

where $C(x)$ is the skew-CDF concentration profile, and x_L and x_R are the left and right limits of x.

The diffusivity of a species is a function of concentration. Thus, as the concentration profile changes with diffusion, the diffusivity also does not remain same throughout the system. This dependence of diffusivity on concentration also results in the non-symmetric nature of concentration profiles across the interface. Once the Matano plane was identified, we calculated the diffusivity as a function of composition using the B-M solution [4, 21] for a semi-infinite slab. At any concentration c, the diffusivity $D(c)$ is obtained by solving the following equation:

$$-\frac{1}{2}\int_{0}^{c} xdC = tD(c)\left[\frac{dC}{dx}\right]_{at\ c}$$

where t is time and x is the position on concentration profile where concentration, $C = c$. The term $\int_0^c xdC$ represents the area under the x vs. C curve from $C = 0$ to $C = c$, as shown in figure. 3b. The term $\left[\frac{dC}{dx}\right]_{at\ c}$ denotes the slope of C vs. x curve at $C = c$. Thus, we obtained the value of diffusivity at difference value of concentration, as shown in Fig. 4, by solving this equation iteratively over increasing values of concentration along the concentration profile curve.

Concentration profiles for Ni, Ti and V diffusion across the interfaces at 1000 °C were extracted after 30 ps and are shown in Fig. 2. The dotted green lines represent the interface at the initial time step and orange lines represent interface after 30 ps. Shift in the interface after diffusion is clearly visible at Ti-V and Ni-V interfaces and the same is negligible in Ni-V interface. However, the atoms of the elements on either side have diffused to the other side in all the cases. Diffusion coefficients were calculated using B-M analysis for which Matano plane is shown in Fig. 3. It is the plane constructed at the intersection of the concentration profiles where the curves are symmetric i.e., the area under the curve on left side is equal to the area above the curve on the right side.

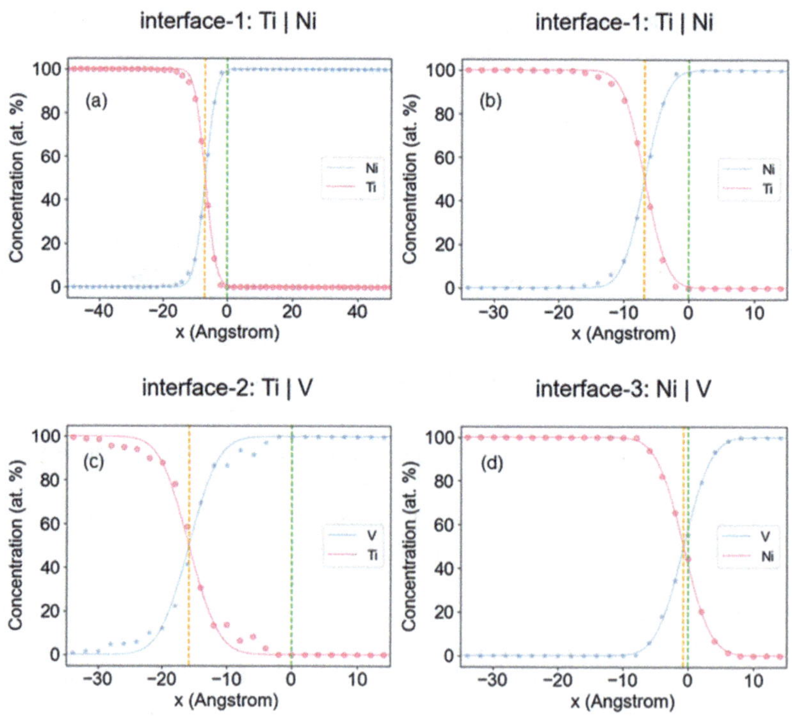

Fig. 2. (a) Complete concentration profile for Ti-Ni interface; selected regions for (b) Ti-Ni (c) Ti-V (d) Ni-V interface after 30 ps

In the extracted concentration profiles at Ti-Ni and Ti-V interface, interfacial shift towards the left was observed indicating preferred movement of Ti atoms towards right. The shift in Matano interface is correlated to the size difference between the diffusing elements. When interdiffusion happens across interface, we would expect the interface to shift towards the element with larger size and the amount of shift should be directly related to the magnitude of the size difference and the amount of diffusion which happens [22]. So, when comparing Ti-Ni and Ti-V interface, Matano interface is expected to shift towards Ti side in both the cases as Ti has larger atomic size than both Ni and Ti. The same is observed in Fig. 2. But the amount of shift observed in Ti-Ni is less than that observed in Ti-V even though the size difference between the Ti-Ni is more as seen in Table 1. This may be attributed to the second factor which is the extent of diffusion which is correlated with the bond dissociation energies. Ni-Ni bond has higher dissociation energy as compared to V-V bond and thus, the diffusion of V is easier than Ni. This may be the reason as to why the shift in Matano interface in case of Ti-V is more as compared to that in Ti-Ni. Similarly in the case of Ni-V interface, we observe negligible shift due to similar sizes and extremely small difference in their bond dissociation energies as evident from Fig. 2. Interdiffusion coefficients for Ti at Ti-Ni interface, Ti at Ti-V interface and Ni at Ni-V interfaces are shown in Fig. 4.

Table 1. Atomic radii and bond dissociation energies of diffusing elements

Element	Atomic radius (Å) [23]	Bond dissociation energy [24]
Ni-Ni	0.135	261.9
Ti-Ti	0.140	141
V-V	0.135	242

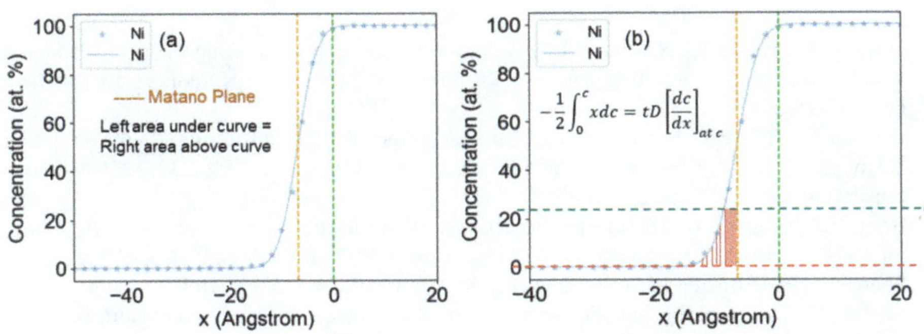

Fig 3. (a) Orange dotted lines represent the Matano plane (b) calculation of diffusion coefficient using B-M analysis

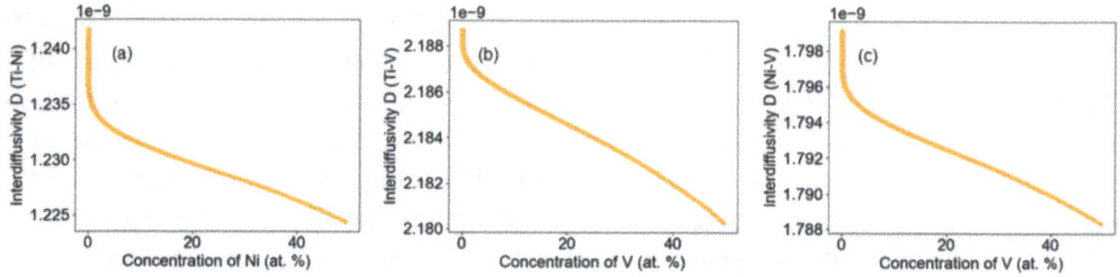

Fig. 4. Diffusion coefficients (in m^2/s) for (a) Ti in Ti-Ni interface (b) Ti in Ti-V interface and (c) Ni in Ni-V interface

4 Conclusion

We have developed a method to calculate the interdiffusion coefficients through atomistic simulations of diffusion multiples. A simulation cell containing Ni-Ti-V diffusion multiple was created and heated at 1000 °C to allow diffusion of elements. The concentration profiles across interfaces were extracted and fitted to skew cumulative distribution functions and subsequently, the interdiffusion coefficients for binary interfaces were calculated as a function of concentration using the Boltzmann-Matano analysis. The shifts in Matano plane corelate well with the size differences and bond dissociation energies of the diffusing elements across the interface.

Acknowledgements. We gratefully acknowledge Dr. Ganesh Balasubramanian for facilitating the use of Hawk cluster at Lehigh University for performing computational studies. Jhalak and DB would like to acknowledge support from Prime Minister's Research Fellowship (PMRF) scheme by Ministry of Education, Government of India.

References

1. Ouyang, G., Ray, P.K., Kramer, M.J., Akinc, M.: Pressureless sintering of Mo-Si-B alloys with Fe additive. J. Mater. Eng. Perform. **26**, 2417–2422 (2017). https://doi.org/10.1007/s11665-017-2657-z
2. Ouyang, G., Ray, P.K., Kramer, M.J., Akinc, M.: Effect of AlN substitutions on the oxidation behavior of ZrB2–SiC composites at 1600 °C. J. Am. Ceram. Soc. **99**, 3389–3397 (2016). https://doi.org/10.1111/jace.14345
3. Borg, R.J., Dienes, G.J.: Solid State Reactions. In: An introduction to solid state diffusion, pp. 287–314. Elsevier (1988). https://doi.org/10.1016/B978-0-12-118425-4.50017-3
4. Shewmon, P.: Diffusion in Solids, 2nd edn. McGraw-Hill, New York (1963)
5. Mehrer, H.: Diffusion: introduction and case studies in metals and binary alloys, In: Heitjans, P. Kärger, J.: Diffus. Condens. Matter Methods Mater. Models, Springer, Berlin, Heidelberg, pp. 3–63 (2005)https://doi.org/10.1007/3-540-30970-5_1
6. Roy, I., Ray, P.K., Balasubramanian, G.: Examining oxidation in β-NiAl and β-NiAl+Hf alloys by stochastic cellular automata simulations. Npj Mater. Degrad. **5**, 1–9 (2021). https://doi.org/10.1038/s41529-021-00202-4
7. Roy, I., Ray, P.K., Balasubramanian, G.: Modeling oxidation of AlCoCrFeNi High-Entropy alloy using stochastic cellular automata. Entropy **24**, 1263 (2022). https://doi.org/10.3390/e24091263

8. Fick, D.A.: On liquid diffusion, Lond. Edinb. Dublin Philos. Mag. J. Sci. 10 (1855) 30–39https://doi.org/10.1080/14786445508641925
9. Roy, I., Ray, P.K., Balasubramanian, G.: Diffusion of multi-principal elements through stable Cr2O3 and Al2O3 scales. Materialia **24**, 101497 (2022). https://doi.org/10.1016/j.mtla.2022.101497
10. Beniwal Jhalak, D., Ray, P.K.: Data-driven phase selection, property prediction and force-field development in multi-principal element alloys, In: Verma, A. Mavinkere Rangappa, S. Ogata, S. Siengchin, S.: Forcefields At.-Scale Simul. Mater. Appl., Springer Nature, Singapore, 2022: pp. 315–347 https://doi.org/10.1007/978-981-19-3092-8_16
11. Yin, X., et al.: Diffusivity and atomic mobility for Fcc Ni–Ti–V alloys: Experiment and modeling. Calphad **82**, 102576 (2023). https://doi.org/10.1016/j.calphad.2023.102576
12. Zhu, L., et al.: Measurement of interdiffusion and impurity diffusion coefficients in the bcc phase of the Ti–X (X = Cr, Hf, Mo, Nb, V, Zr) binary systems using diffusion multiples. J. Mater. Sci. **52**, 3255–3268 (2017). https://doi.org/10.1007/s10853-016-0614-0
13. Zhao, J.C.: Diffusion multiple screening: phase diagram determination and related studies, In: Buschow, K.H.J., Eds.:Encycl. Mater. Sci. Technol., Elsevier, Oxford, pp. 1–13 (2005) https://doi.org/10.1016/B0-08-043152-6/02041-6
14. Zhao, J.-C.: The Diffusion-Multiple Approach to Designing Alloys. Annu. Rev. Mater. Res. **35**(1), 51–73 (2005). https://doi.org/10.1146/annurev.matsci.35.100303.111314
15. Zhao, J.-C., Zheng, X., Cahill, D.G.: High-throughput diffusion multiples. Mater. Today **8**, 28–37 (2005). https://doi.org/10.1016/S1369-7021(05)71122-6
16. Thompson, A.P.: LAMMPS - a flexible simulation tool for particle-based materials modeling at the atomic, meso, and continuum scales. Comput. Phys. Commun. **271**, 108171 (2022). https://doi.org/10.1016/j.cpc.2021.108171
17. Hirel, P.: Atomsk: a tool for manipulating and converting atomic data files. Comput. Phys. Commun. **197**, 212–219 (2015). https://doi.org/10.1016/j.cpc.2015.07.012
18. Maisel, S.B., Ko, W.-S., Zhang, J.-L., Grabowski, B., Neugebauer, J.: Thermomechanical response of NiTi shape-memory nanoprecipitates in TiV alloys. Phys. Rev. Mater. **1**, 033610 (2017). https://doi.org/10.1103/PhysRevMaterials.1.033610
19. O'hagan, A., Leonard, T.: Bayes estimation subject to uncertainty about parameter constraints. Biometrika **63**(1), 201–203 (1976). https://doi.org/10.1093/biomet/63.1.201
20. Owen, D.B.: A table of normal integrals. Commun. Stat. - Simul. Comput. **9**, 389–419 (1980). https://doi.org/10.1080/03610918008812164
21. Matano, C.: On the relation between the diffusion-coefficients and concentrations of solid metals. Jpn. J Phys **8**, 109 (1933)
22. Jhalak, G., Balasubramanian, P.K.: Ray, effect of HF and AI on self-diffusion in amorphous silica using molecular dynamics. JOM (2023). https://doi.org/10.1007/s11837-023-06323-2
23. WebElements Periodic Table » Periodicity » Radius - metallic (12) » Periodic table gallery, (n.d.). https://www.webelements.com/periodicity/metallic_radius/ (Accessed 14 Jan 2024)
24. B. deB. Darwent, Bond dissociation energies in simple molecules, (1970). https://srd.nist.gov/NSRDS/NSRDS-NBS31.pdf
25. Author, F.: Article title. Journal **2**(5), 99–110 (2016)
26. Author, F., Author, S.: Title of a proceedings paper. In: Editor, F., Editor, S. (eds.) CONFERENCE 2016, LNCS, vol. 9999, pp. 1–13. Springer, Heidelberg (2016).https://doi.org/10.1007/978-3-319-96065-4_6
27. Author, F., Author, S., Author, T.: Book title, 2nd edn. Publisher, Location (1999)
28. Author, F.: Contribution title. In: 9th International Proceedings on Proceedings, pp. 1–2. Publisher, Location (2010)
29. LNCS Homepage, http://www.springer.com/lncs, Accessed 21 Nov 2016

Functional Materials

Analyzing the Mechanical Properties of Shape Memory Nanocomposites Enhanced by Introducing GnP

Ritesh Gupta(✉), Gaurav Mittal, Krishna Kumar, and Upender Pandel

Department of Metallurgical and Materials Engineering, MNIT Jaipur, JLN Marg, Malviya Nagar, Jaipur, Rajasthan 302017, India
ritesh.gupta641@gmail.com

Abstract. Shape memory polymers (SMPs) are capable of enduring significant deformations and returning to their original form upon activation by certain external stimuli. However, their restricted mechanical capabilities have limited their broader application in engineering fields. To address this, the integration of graphene nanoplatelets (GnPs) with SMPs has proven effective in enhancing their mechanical properties while maintaining inherent shape memory functions. The study evaluated shape memory nanocomposites (SMNCs) using static tensile, flexural and shape memory tests, along with scanning electron microscopy to analyze tensile fractures. The findings indicated that the SMNCs achieved their best shape memory and mechanical performance at a GnP concentration of 0.6 wt%.

Keywords: Shape memory nanocomposites · graphene nanoplatelets · EMT model

1 Introduction

Shape memory materials (SMMs) stand out for their extraordinary capability to react to external triggers such as temperature, moisture, magnetic fields, and pH levels, enabling them to transform between various shapes dynamically. These materials are primarily divided into three categories: shape memory polymers (SMPs), shape memory alloys (SMAs), and shape memory foams (SMFs). They have found diverse applications across several fields, including smart structures, space deployment mechanisms that can be collapsed, artificial arteries and muscles, bendable electronics, and in the domains of energy storage and conversion [1–6]. Relative to SMAs, SMPs offer superior qualities including high capacity for shape deformation, tolerance to stress, biodegradability, ease of manufacturing and processing, along with excellent shape recovery. These attributes, combined with their lighter weight and cost efficiency, make SMPs a more suitable option for diverse applications, particularly in biomedical and structural fields. Nevertheless, SMPs do face limitations regarding their mechanical strength, a challenge that can be effectively mitigated by incorporating nanofillers [7]. For this purpose, epoxy resin

A. Tiwari et al. (Eds.): iConFIRM 2023, SPPHY 308, pp. 151–164, 2024.
https://doi.org/10.1007/978-981-97-4557-9_16

was selected as the foundational material, owing to its advantageous mechanical and thermomechanical properties, resistance to chemical and thermal deterioration, and cost-effectiveness [8, 9].

GnPs were selected as nanofillers due to their multifunctionality and high aspect ratio, significantly bolstering the mechanical properties of structural adhesives and facilitating effective load transfer between the polymer matrix and the nanofillers. This leads to the production of exceptional SMNCs [10–14]. Compared to multi-walled carbon nanotubes (MWCNTs), GnPs offer several key advantages as nanofillers. They enhance the mechanical properties of composites more effectively, thanks to their extensive surface area and superior interaction with matrix materials [15]. Their two-dimensional structure provides improved thermal conductivity, suitable for thermal management applications [16]. In polymer applications, GnPs offer better gas or moisture barrier resistance than MWCNTs [17]. Moreover, their 2D structure aids in easier dispersion across various matrices [18]. As manufacturing techniques have evolved, producing GnPs has become more cost-effective, potentially offering economic advantages over MWCNTs in certain scenarios [19]. However, GnPs face specific hurdles when set against MWCNTs. One primary concern is the distribution of GnPs in the polymer matrix. Even though the two-dimensional structure of GnPs might seem ideal for dispersion, their tendency to cluster because of strong π-π attractions can lead to inconsistent attributes in the final nanocomposites [20]. Moreover, the modification of GnPs, which is essential for aligning with diverse matrices, can be complex. This modification might compromise its natural characteristics, making it a more intricate task than comparable processes for MWCNTs [19, 21, 22]. To address these challenges, techniques like ultrasonication and magnetic stirring have been adopted to ensure uniform dispersion of GnPs within the polymer matrix. In this research, weight concentrations of GnPs at 0.4%, 0.6%, and 0.8% were chosen to optimize dispersion and amplify the mechanical properties of the nanocomposites. Maintaining this deliberate control over GnPs distribution is pivotal to harnessing its full potential.

2 Materials

For the creation of SMPs test specimens, an aromatic epoxy resin and aliphatic amine hardener were sourced from Huntsman International (India) Private Limited. The epoxy resin, chemically named Diglycidyl Ether of Bisphenol A (DGEBA), and the aliphatic amine, Triethylenetetramine (TETA), were available commercially as LY556 and HY951, respectively. The nanofiller GnPs, having a thickness ranging between 1–2 nm, were procured from Ad-Nano Technologies Private Limited (India).

3 Preparation of Epoxy/GnP SMNCs

GnPs underwent 5 h heating process in a vacuum oven at 393 K to remove any remaining moisture. These GnPs were then dispersed in a 100 ml acetone solution, with extensive stirring for an hour using a magnetic stirrer. The addition of acetone was key in ensuring a uniform distribution of the nanofillers when subsequently mixed with epoxy resin, thereby improving the dispersion of the GnPs within the matrix. To aid in degassing, the epoxy was subjected to a vacuum degassing procedure for one hour, preparing it for subsequent processing steps. After degassing, the epoxy resin was mixed with the acetone and nanoparticle solution, followed by a 30-min ultrasonication to enhance the homogeneity of mixture. The ultrasonication process introduced air bubbles, which were subsequently removed by placing the mixture in a vacuum chamber for 6 h at 353 K, preventing voids in the final nanocomposites. For polymerization, an aliphatic amine-based hardener was added to the resin in a 10:1 weight ratio. After thorough mixing, the solution was poured into Teflon molds and left to cure at room temperature for 24 h. To further ensure the elimination of air pockets, a post-curing process was conducted in a vacuum oven at 353 K for 6 h, in line with the guidelines provided by the supplier and literature references. The cured nanocomposite plates were then precisely shaped into samples using abrasive water jet machining. The study experimented with various GnP concentrations, specifically 0.4%, 0.6%, and 0.8% by weight.

4 Material Characterization

Tensile properties of the nanocomposites were evaluated using an Instron 5967 tensile testing machine with a ±30 kN load cell, following ASTM D638 standards at a crosshead speed of 1 mm/min. Flexural strength was tested using the same machine, equipped with a ±5 kN load cell and conducted at 1 mm/min, adhering to ASTM D790 guidelines. The setup maintained a 16:1 span-to-depth ratio. For each type of nanocomposite, three independent experimental trials were conducted for the tests mentioned above. The average values derived from these trials were precisely recorded and later reported, ensuring accuracy and reliability in the findings. FESEM analysis, using a Nova NanoSEM 450, was performed on samples fractured during tensile tests. This analysis aimed to examine the interfacial bonding between the epoxy matrix and GnPs, assess the influence of GnP incorporation on stress transfer, and observe the dispersion and agglomeration of GnPs within the matrix. The findings were thoroughly analyzed and discussed.

5 Shape Memory Analysis

Three samples of each configuration were used in heat-activated bending tests to evaluate the shape memory properties of both neat epoxy and epoxy/GnP SMNCs. Specimens, sized 100 × 10 × 3 mm^3, were heated from room temperature to 360 K at 5 K/min and held at 360 K for 5 min to reach the rubbery stage. During the programming stage, they were bent to a 90° angle (θ_{max}) and held for 5-min. After cooling to room temperature at 5 K/min, a minor elastic recovery occurred, resulting in an angle θ_{fix}, indicating shape fixity. In the recovery stage, reheating to 360 K caused a change in bending angle, denoted as θ_{remain}. Shape recovery was evaluated from these observations and calculated using Eqs. (1)–(3) [23],

$$\text{Shape fixity ratio}\,(\theta_{SFR}) = \frac{\theta_{fix}}{\theta_{max}} \times 100\% \tag{1}$$

$$\text{Shape recovery ratio}\,(\theta_{SRR}) = \frac{\theta_{fix} - \theta_{remain}}{\theta_{max}} \times 100\% \tag{2}$$

$$\text{Shape recovery rate } (\theta_{rate}) = \frac{\pi(\theta_{fix} - \theta_{remain})}{180t} \tag{3}$$

The entire heat activation test was documented using an image capture system, which included a high-resolution camera. The captured images were then processed in AutoCAD to assess the bending angles over time during the recovery stage of the shape memory cycle.

6 Results and Discussion

6.1 Tensile Testing

Figure 1(a) displays the tensile stress-strain curves for neat epoxy SMP and epoxy/GnP SMNCs with different concentrations of GnPs. Figure 1(b) illustrates the effect of GnP concentration on the Young's modulus and ultimate tensile strength (UTS) of the SMNCs, including error bars to indicate the maximum and minimum values observed. The neat epoxy SMP exhibited a tensile strength of about 35 MPa and a Young's modulus of roughly 3.3 GPa. Incorporating GnPs into the epoxy matrix led to notable enhancements in tensile strength. Specifically, with GnP concentrations at 0.4 wt% and 0.6 wt%, the tensile strength increased to approximately 47.4 MPa and 61.42 MPa, respectively. These values represent increases of 33.14% and 72.52% compared to neat epoxy. Similarly, Young's modulus showed a significant rise, reaching around 4.22 GPa for 0.4 wt% GnP and 4.54 GPa for 0.6 wt% GnP, amounting to increases of 27.87% and 37.57%, respectively, when compared to neat epoxy. Nonetheless, with the increase in GnP concentration to 0.8 wt% in the epoxy matrix, there was a noticeable decrease in both tensile strength and modulus, which dropped to 57.33 MPa and 4.37 GPa, respectively. This suggests that there exists an optimal concentration range (0.4 wt% to 0.6 wt%) of GnPs that results in the highest tensile strength and modulus in the SMNCs. Furthermore, the strain at UTS (%) decreased with the increasing concentration of GnPs in the epoxy matrix. Experimental results showed increased Young's modulus with 0.4 to 0.6 wt% GnPs, but a decrease at 0.8 wt%, possibly due to agglomeration at higher filler content. This suggests that substantial GnPs agglomeration can act as a stress concentrator, negatively affecting the mechanical properties of the SMNCs. This agglomeration effect is comprehensively analyzed through the EMT model in section 0, while morphological studies will be discussed in Sect. 6.3.

Micromechanical Model of Epoxy/GnP SMNCs. In this section, the micromechanics of the two-phase SMNCs hybridized with GnPs are explored to validate the experimental results with the theoretical mathematical model. It is vital to recognize the tendency of nanosized reinforcing phases, such as GnPs, to agglomerate, leading to a nonuniform filler distribution within nanocomposites. GnPs are believed to be straight and uniformly spread within the interphase. Nevertheless, once the local volume fraction of GnPs reaches a certain limit, the agglomeration of GnPs might become significant due to Van der Waals forces. At this point, interactions between GnPs cannot be ignored. In the production process, certain imperfections are inevitable. The current methodology does not account for manufacturing defects like voids, cracks, and delamination. Subsequent enhancements to this model should focus on these challenges.

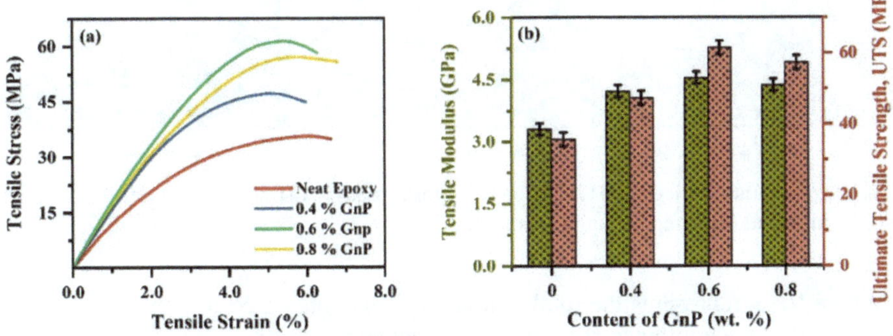

Fig. 1. (a) Tensile stress–strain curves of neat epoxy and three different epoxy/GnP SMNCs and (b) variation of tensile Young's modulus and ultimate tensile strength of neat epoxy SMP and SMNCs concerning different GnP concentration.

In this study, the Eshelby–Mori–Tanaka (EMT) model [24, 26] was utilized to evaluate the mechanical properties of the hybrid matrix enriched with GnPs. This model was chosen for its ability to incorporate additional factors like nanoparticle agglomeration. To delve into the theoretical exploration of this agglomeration phenomenon, this hypothesis involved specific GnPs aggregating within spherical regions in the matrix, while the remaining ones maintained a uniform dispersion. In this scenario, the matrix was considered isotropic and characterized by Young's modulus (E_m), and Poisson's ratio (ν_m). Consequently, the composite was divided into two distinct segments: the inclusion phase, encompassing areas with concentrated reinforcement, and the effective matrix phase, representing regions with a uniform filler dispersion. It is crucial to emphasize that in both phases, GnPs were assumed randomly oriented. In this model, a two-parameter description was employed to capture this distinct distribution, with the agglomeration parameters expressed in Eqs. (4) and (5). Parameter μ signifies the volume fraction of the inclusion phase $\left(V_{ip}\right)$ over the total representative element volume (V), and η, represents the volume fraction of GnPs within the inclusion phase $\left(V_{GnP}^{ip}\right)$ over the total GnP volume (V_{GnP}). Three distinct cases arise based on these parameters. In the first scenario ($\eta = \mu = 1$), GnPs are evenly spread throughout the matrix, indicating no agglomeration, as depicted in Fig. 2(a). The second case involves partial GnPs agglomeration, where $\eta > \mu$ and $\mu < 1$, as demonstrated in Fig. 2(b). In the third case ($\eta > \mu$ and $\eta = 1$) and, illustrated in Fig. 2(c), all GnPs are concentrated within the inclusion phase.

$$\mu = \frac{V_{ip}}{V} \tag{4}$$

$$\eta = \frac{V_{GnP}^{ip}}{V_{GnP}} \tag{5}$$

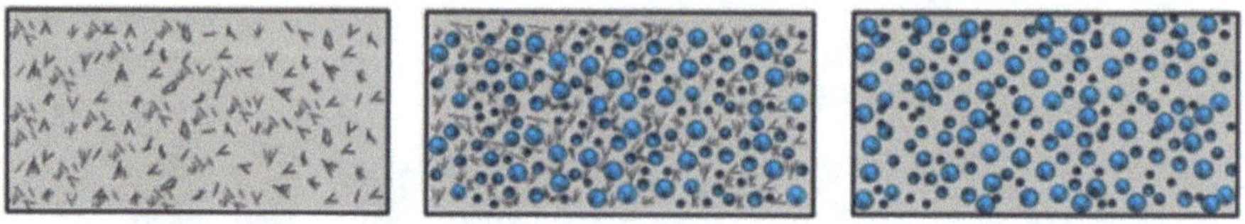

(a) $\eta = \mu = 1$ (b) $\eta > \mu$ and $\mu < 1$ (c) $\eta > \mu$ and $\eta = 1$

Fig. 2. Agglomeration model of GnPs for the hybrid matrix: **(a)** no agglomeration, **(b)** partial agglomeration and **(c)** complete agglomeration.

Where V_{GnP} represents the total volume of GnPs integrated into the matrix, it can be described as the combined sum of two distinct contributions,

$$V_{GnP} = V_{GnP}^{em} + V_{GnP}^{ip} \tag{6}$$

In this scenario, where V_{GnP}^{ip} represents the volume of GnPs within the inclusion phase and V_{GnP}^{em} represents the volume of GnPs scattered in the effective matrix phase and the total volume (V) of a representative element can be expressed using the following relationship,

$$V = V_m + V_{GnP} \tag{7}$$

Where, V_m represents the volume of the polymer matrix. Typically, the volume fraction of GnPs (ϑ_{GnP}), can be established as a function of the corresponding weight fraction (w_{GnP}),

$$\vartheta_{GnP} = \frac{1}{\left(\frac{\rho_{GnP}}{\rho_m w_{GnP}} - \frac{\rho_{GnP}}{\rho_m} + 1\right)} \tag{8}$$

The density of matrix and GnPs is represented by ρ_m and ρ_{GnP}, respectively. The mechanical properties of the polymer matrix, specifically its engineering constants, the E_m and the G_m, are 3.30 GPa and 1.18 GPa respectively, as obtained from reference [27]. Additionally, the elastic properties of GnPs are presented in Table 1 in terms of Hill's elastic moduli derived from [28, 29].

Table 1. Mechanical properties of GnPs in form of Hill's elastic moduli.

Material	k_{GnP} (GPa)	l_{GnP} (GPa)	p_{GnP} (GPa)	n_{GnP} (GPa)	m_{GnP} (GPa)	ρ_{GnP} (gm/cm^3)
GnP	850	6.8	102000	102000	369	2.2

The bulk modulus and shear modulus of the inclusion phase $\left(K_{ip}^*, G_{ip}^*\right)$ and the effective matrix phase $\left(K_{em}^*, G_{em}^*\right)$ can be determined as follows,

$$K_{ip}^* = K_m + \frac{\vartheta_{GnP}\eta(\delta - 3K_m\alpha)}{3(\mu - \vartheta_{GnP}\eta + \vartheta_{GnP}\eta\alpha)} \tag{9}$$

$$G^*_{ip} = G_m + \frac{\vartheta_{GnP}\eta(\gamma - 2G_m\beta)}{2(\mu - \vartheta_{GnP}\eta + \vartheta_{GnP}\eta\beta)} \quad (10)$$

$$K^*_{em} = K_m + \frac{\vartheta_{GnP}(1 - \eta)(\delta - 3K_m\alpha)}{3(1 - \mu - \vartheta_{GnP}(1 - \eta) + \vartheta_{GnP}(1 - \eta)\alpha)} \quad (11)$$

$$G^*_{em} = G_m + \frac{\vartheta_{GnP}(1 - \eta)(\gamma - 2G_m\beta)}{2(1 - \mu - \vartheta_{GnP}(1 - \eta) + \vartheta_{GnP}(1 - \eta)\beta)} \quad (12)$$

For clarity, it is crucial to mention that the superscript "*" is utilized to denote properties related to the hybrid matrix, and subscripts "ip" and "em" are used to distinguish the inclusion phase and effective matrix phase, respectively. Moreover, α, β, δ, and γ are defined as follows,

$$\alpha = \frac{3K_m + 2n_{GnP} - 2l_{GnP}}{3n_{GnP}} \quad (13)$$

$$\beta = \frac{4G_m + 7n_{GnP} + 2l_{GnP}}{15n_{GnP}} + \frac{2G_m}{5p_{GnP}} \quad (14)$$

$$\delta = \frac{3K_m(n_{GnP} + 2l_{GnP}) + 4\left(k_{GnP}n_{GnP} - l^2_{GnP}\right)}{3n_{GnP}} \quad (15)$$

$$\gamma = \frac{2}{15}\left(k_{GnP} + 6m_{GnP} + 6G_m - \frac{2G_m l_{GnP} + l^2_{GnP}}{n_{GnP}}\right) \quad (16)$$

Where, k_{GnP}, l_{GnP}, p_{GnP}, n_{GnP} and m_{GnP} represent Hill's elastic moduli of the GnPs, determined by aligning Hill's elastic matrix with the elastic stiffness matrix for GnPs. The hybrid matrix's effective bulk modulus and shear modulus, incorporating GnPs can be calculated as follows,

$$K^*_m = K^*_{em}\left(1 + \frac{\mu\left(\frac{K^*_{ip}}{K^*_{em}} - 1\right)}{1 + (1 - \mu)\left(\frac{K^*_{ip}}{K^*_{em}} - 1\right)\left(\frac{1+\nu^*_{em}}{3-3\nu^*_{em}}\right)}\right) \quad (17)$$

$$G^*_m = G^*_{em}\left(1 + \frac{\mu\left(\frac{G^*_{ip}}{G^*_{em}} - 1\right)}{1 + (1 - \mu)\left(\frac{G^*_{ip}}{G^*_{em}} - 1\right)\left(\frac{8-10\nu^*_{em}}{15-15\nu^*_{em}}\right)}\right) \quad (18)$$

Here, it represents the Poisson's ratio of the effective matrix phase and is defined as follows,

$$\nu^*_{em} = \frac{3K^*_{em} - 2G^*_{em}}{6K^*_{em} + 2G^*_{em}} \quad (19)$$

The properties of the isotropic hybrid matrix, including the Young's modulus $\left(E^*_m\right)$ and Poisson's ratio $\left(\nu^*_m\right)$, can be calculated as follows,

$$E^*_m = \frac{9K^*_m G^*_m}{3K^*_m + G^*_m} \quad (20)$$

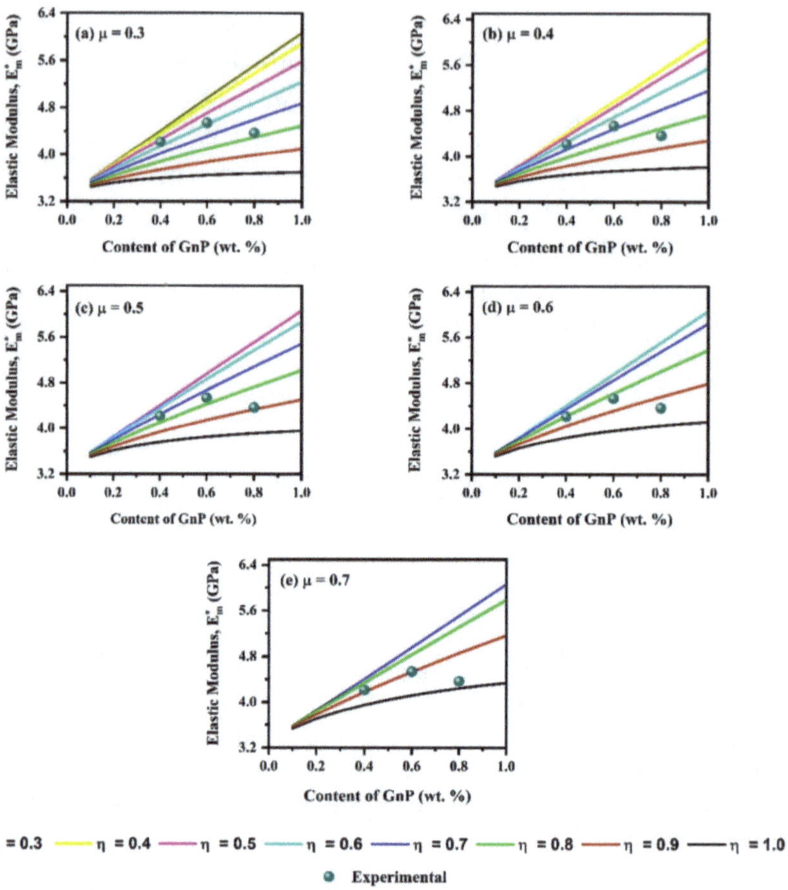

Fig. 3. The effective Young's modulus obtained from EMT approach with volume fraction of GnPs at different levels of agglomeration when inclusion phase size is (a) $\mu = 0.3$, (b) $\mu = 0.4$, (c) $\mu = 0.5$, (d) $\mu = 0.6$ and (e) $\mu = 0.7$.

$$\nu_m^* = \frac{3K_m^* - 2G_m^*}{6K_m^* + 2G_m^*} \tag{21}$$

As mentioned earlier, the parameter μ represents the volume of the inclusion phase (or the size of the inclusion phase), and the parameter η represents the volume of GnPs inside the inclusion phase. To ensure clarity, the volume of the inclusion phase is initially assumed to be in the range of 30 to 70% ($\mu = 0.3$ to 0.7), and the parameter η varies with the weight percentage of GnPs. The results of the mathematical simulation are compared with experimental data as shown in Fig. 3(a–e). As the agglomeration parameter η increases, the theoretical effective Young's modulus decreases, indicating that greater GnPs agglomeration leads to a reduction in modulus. Additionally, a higher value of μ suggests a more uneven distribution of GnPs within the matrix. Although μ values ranging from 0.1 to 0.9 were considered in the study, for clarity, only results for

μ values from 0.3 to 0.7 are presented. Figure 3(a–e) show that the experimental elastic modulus for 0.4 and 0.6 wt% GnPs concentrations aligns with the modulus predicted by the EMT model for $\mu = 0.7$ and $\eta = 0.9$. This implies that about 70% of the volume is part of the inclusion phase, with approximately 90% of GnPs contained within this phase. Conversely, for 0.8 wt% GnP (as seen in Fig. 3e), the experimental elastic modulus falls between $\eta = 0.9$ *to* 1.0, signifying that 90 to 100% of GnPs are within the inclusion phase, indicating increased agglomeration at this concentration, which in turn lowers the elastic modulus. These findings suggest that the EMT model effectively predicts the influence of GnPs agglomerations on the elastic properties. The accuracy of these conclusions about GnPs dispersion will be further corroborated by FESEM micrographs, as discussed in Sect. 6.3.

6.2 Flexural Testing

Figure 4a displays the flexural stress-strain curves for epoxy resin composite samples with different GnP weight fractions, tested under room temperature conditions. Additionally, Fig. 4b presents the flexural strength and modulus measurements for neat epoxy SMP and the three different epoxy/GnP SMNCs with varying GnP contents, all recorded at room temperature. Room temperature testing of composite samples with 0 wt%, 0.4 wt%, 0.6 wt%, and 0.8 wt% GnPs revealed flexural strengths of 63.02 MPa, 74.28 MPa, 116.37 MPa, and 93.30 MPa, respectively, with flexural moduli of 2.38 GPa, 2.75 GPa, 3.65 GPa, and 3.10 GPa. Thus, relative to neat epoxy SMP, the flexural strengths for 0.4 wt%, 0.6 wt%, and 0.8 wt% GnP composites were found to be greater by 17.86%, 84.65%, and 48.04%, respectively. Likewise, the flexural modulus showed increments of 19.56%, 58.69%, and 34.78% for the respective GnP concentrations. Among these, the composite with 0.6 wt% GnP exhibited the highest flexural strength and modulus. However, a decrease in flexural properties was observed when the GnPs content went beyond 0.6 wt%. This reduction in performance is mainly due to the inhomogeneity introduced in the epoxy SMP cured product network by the addition of GnPs, resulting in a micro nonuniform continuous structure that compromised the material strength. Overall, incorporating GnPs into the epoxy SMP significantly improved its flexural strength, with a more notable enhancement evident in three-point bending tests compared to static tensile testing.

6.3 Morphological Analysis

A thorough analysis of the tensile properties of both neat epoxy SMP and epoxy/GnPs SMNCs was conducted by observing the tensile fracture surfaces through FESEM. The neat epoxy SMP displayed a comparatively smooth fracture surface as seen in Fig. 5a, which contrasts with the coarser textures of the epoxy/GnP SMNCs shown in Fig. 5b, c, and d. The roughness of these surfaces is attributed to the unevenly distributed GnPs, some of which are visible as white clusters on the surface, while others are embedded within the epoxy matrix. Notably, the size of these clusters tended to increase with the concentration of GnPs, suggesting a tendency for the nanoplatelets to agglomerate. The enhancements in the mechanical characteristics of the epoxy/GnP SMNCs from multiple

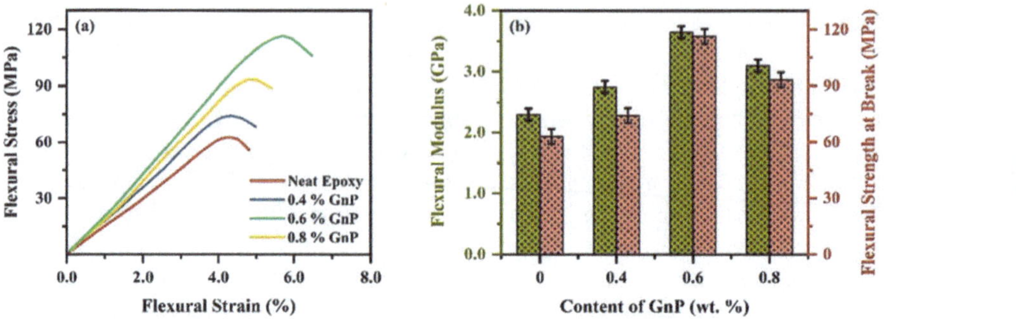

Fig. 4. (a) Flexural stress–strain curves of neat epoxy SMP and three different epoxy/GnP SMNCs and (b) variation of flexural Young's modulus and flexural strength at break of SMNCs concerning different GnP concentration.

contributing factors. First, the uniformly distributed GnPs increase the interfacial contact area within the epoxy matrix, improving the transfer of stress from the epoxy to the GnPs. Furthermore, the two-dimensional form of GnPs results in a high aspect ratio, enlarging their contact surface with the matrix. Such an increase in the interfacial area due to the shape of GnPs enhances the load transfer efficiency, which in turn boosts the mechanical robustness of SMNCs. Additionally, the formation of a durable interface between the GnPs and the epoxy matrix is crucial. This robust interface secures the GnPs within the matrix, mitigating any detachment under mechanical stress. Notably, at GnP concentrations up to 0.6 wt%, there are strong indications of superior wetting, which reflects a strong interfacial adhesion and efficient load transfer dynamics. These elements work in concert to substantially enhance the mechanical strength of the SMNCs.

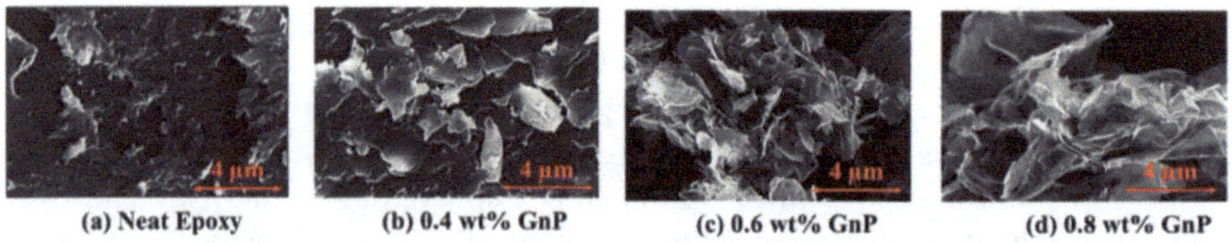

Fig. 5. FESEM micrographs of **(a)** neat epoxy SMP and epoxy/GnPs SMNCs with **(b)** 0.4 wt% GnP, **(c)** 0.6 wt% GnP and **(d)** 0.8 wt% GnP.

However, it is important to note that when the content of GnP exceeds 0.6 wt%, a downturn in the mechanical properties begins to emerge. This decline at higher GnPs concentrations can be linked to a number of factors. To start with, the clustering of GnPs bundles, attributed to their π-π interactions, tends to increase the viscosity of the epoxy, which can act as a precursor to crack formation when the material is under stress. Additionally, the challenge of fully encapsulating the GnPs with the epoxy owing to the dramatic rise in the surface area to volume ratio of filler can lead to increased porosity and the creation of defect sites within the nanocomposites, thereby weakening the mechanical integrity. Finally, a reduction in the cross-linking density within the nanocomposites may become more prevalent due to the strong interface and even spread of GnPs, highlighting the critical role that both interface strength and dispersion play in dictating the mechanical performance of epoxy/GnP SMNCs.

6.4 Shape Memory Behavior Analysis

In Fig. 6(a), the experimental results of the bending angle recovery for neat epoxy SMP and three SMNCs cantilever beam samples are presented. Complete recovery for neat epoxy SMP occurred in 100 s, whereas epoxy/GnP SMNCs, achieved full recovery in less than 100 s. The incorporation of GnPs in the epoxy matrix improved the thermal response of SMNCs, reducing the recovery time by 21%, 52%, and 35%, for 0.4 wt%, 0.6 wt%, and 0.8 wt% GnP configuration, respectively. However, in nanocomposite with 0.8 wt% GnP, a recovery delay similar to neat epoxy SMP was observed due to nanoparticle agglomeration within the matrix, leading to a decrease in modulus. The shape memory parameters including the θ_{SFR}, θ_{SRR} and θ_{rate}, were calculated using Eqs. (1)–(3) as described in Sect. 0. Among these parameters, θ_{SFR} and θ_{SRR} are visually represented in Fig. 6(b). The shape fixity capabilities of neat epoxy SMP are exemplified by θ_{SFR} of 98.77%, whereas this ratio shows a slight decrease with other epoxy/GnP SMNCs, registering at 97.44% for 0.4 wt% GnP, 95.22% for 0.6 wt% GnP, and 93.11% for 0.8 wt% GnP. In parallel, the θ_{SRR} also exhibit a reduction from the 97.66% of the neat epoxy to 96.33%, 94.11%, and 92.5% for SMNCs with 0.4 wt%, 0.6 wt%, and 0.8 wt% GnP, respectively. The θ_{rate} initially rises with the inclusion of GnPs, peaking at 2.83% for 0.6 wt% GnP, and then falls to 2.07% at 0.8 wt% GnP concentration, likely due to the agglomeration effect of the GnPs. Despite these variations, it is notable that the shape recovery and fixity metrics for all epoxy/GnP SMNCs samples remain commendably above the 90% threshold [23]. Overall, the incorporation of increasing amounts of GnPs in the composite led to restricted movement of the polymer molecular chains, which in turn heightened resistance during the recovery phase of composite. This effect caused a reduction in both the θ_{SRR} and θ_{SFR}. Nevertheless, this addition of GnPs only minimally impacted the shape memory performance of the matrix, highlighting the robust shape recovery capabilities inherent in the SMNCs.

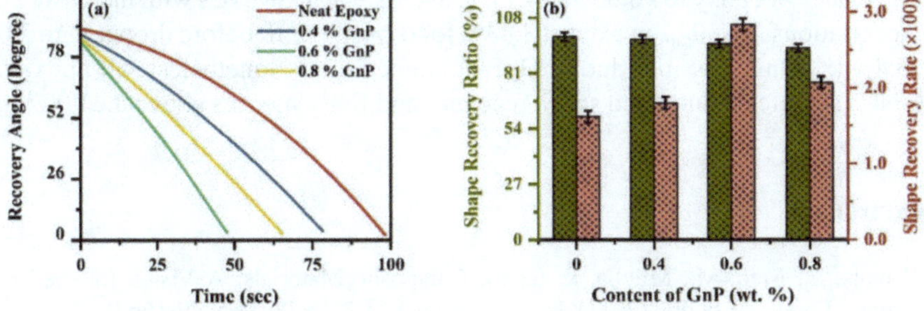

Fig. 6. **(a)** Experimental result of the shape recovery angle vs time of cantilever beam samples made of neat epoxy SMP and three different epoxy/GnP SMNCs and **(b)** variation of θ_{SRR} and θ_{rate} of SMNCs concerning GnP concentrations.

7 Conclusion

In this research, epoxy/GnP SMNCs were successfully developed, and their mechanical and shape memory characteristics, along with those of neat epoxy, were thoroughly examined. The evaluation involved a range of tests, such as static tensile, flexural and shape memory tests. The research led to the following conclusions:

- The GnPs led to significant improvements in the tensile modulus of the SMNCs, with increases of 27.87% for 0.4 wt% GnP and 37.57% for 0.6 wt% GnPs compared to neat epoxy. Even at a higher concentration of 0.8 wt% GnP, the elastic modulus was still greater than that of neat epoxy SMP. The EMT model effectively predicted GnP agglomeration, as the experimental elastic modulus for 0.4 wt% and 0.6 wt% GnP SMNCs corresponded with the EMT model elastic modulus calculated for specific parameters ($\mu = 0.7$ *and* $\eta = 0.9$). This indicated that about 70% of the volume was in the inclusion phase, with approximately 90% of the GnPs embedded in it. Conversely, for 0.8 wt% GnPs, the experimental modulus suggested that between 90 to 100% of the GnPs were in the inclusion phase, pointing to increased agglomeration at higher GnP content, which in turn reduced the modulus. This trend was supported by FESEM micrographs, which showed a smooth surface for the fractured neat epoxy SMP, whereas the epoxy/GnP SMNCs had a more rugged appearance.
- The increase in GnP content led to enhanced flexural properties in the SMNCs. For SMNCs with GnP concentrations of 0.4 wt%, 0.6 wt%, and 0.8 wt%, the flexural strengths were higher than that of neat epoxy SMP by 17.86%, 84.65%, and 48.04%, respectively, at room temperature. Additionally, the flexural modulus saw increments of 19.56%, 58.69%, and 34.78% across these GnP concentrations. Notably, the sample with 0.6 wt% GnP exhibited the highest flexural strength and modulus among the tested samples, recording values of 116.37 MPa and 3.65 GPa, respectively.
- Heat-activated bending tests revealed that adding GnPs to epoxy reduced recovery times by 21%, 52%, and 35% for concentrations of 0.4 wt%, 0.6 wt%, and 0.8 wt% GnPs, respectively. The θ_{SFR} of neat epoxy SMP was 98.77%, which slightly decreased for epoxy/GnP SMNCs, registering at 97.44%, 95.22%, and 93.11% for 0.4 wt%, 0.6 wt%, and 0.8 wt% GnP, respectively. Similarly, the θ_{SRR} also decreased from 97.66% in neat epoxy to 96.33%, 94.11%, and 92.5% in SMNCs with increasing GnP concentrations. The θ_{rate} peaked at 2.83% for 0.6 wt% GnP before dropping to 2.07% at 0.8 wt% GnP, possibly due to GnP agglomeration. Nonetheless, all epoxy/GnP SMNCs samples maintained shape recovery and fixity metrics above the 90% mark.

References

1. Nicolais, L., Meo, M., Milella, E. (eds.): Composite Materials: A Vision for the Future. Springer London, London (2011). https://doi.org/10.1007/978-0-85729-166-0
2. Bergstrom, J.S.: Mechanics of Solid Polymers: Theory and Computational Modeling
3. Arun, D.I., Chakravarthy, P., Arockiakumar, R., Santhosh, B.: Shape Memory Materials, 1st edn. CRC Press/Taylor & Francis Group, Boca Raton, FL. A CRC title, part of the Taylor & Francis imprint, a member of the Taylor & Francis Group, the academic division of T&F Informa plc. CRC Press (2018). https://doi.org/10.1201/9781351119948

4. Parameswaranpillai, J., Siengchin, S., George, J.J., Jose, S. (eds.) Shape memory polymers, blends and composites: advances and applications. In: Advanced Structured Materials, vol. 115. Springer Singapore, Singapore (2020). https://doi.org/10.1007/978-981-13-8574-2
5. Hu, J.: Shape memory polymers: fundamentals, advances and applications. Smithers Rapra, Shawbury (2014)
6. Leng, J., Du, S. (eds.): Shape-memory polymers and multifunctional composites. CRC Press/Taylor & Francis, Boca Raton (2010)
7. Cassatt, J.C., Marini, C.P., Bender, J.W.: The reversible reduction of horse metmyoglobin by the iron(II) complex of trans-1,2-diaminocyclohexane-N, N, N, n-tetraacetate. Biochemistry **14**(25), 5470–5475 (1975). https://doi.org/10.1021/bi00696a014
8. Wilson, J.R., Dicara, L.V.: Effects of previous curare-immobilization on Pavlovian conditioned heart decelerations in the curarized rat. Physiol. Behav. **14**(3), 259–264 (1975). https://doi.org/10.1016/0031-9384(75)90031-1
9. Gasaway, W.C.: Volatile fatty acids and metabolizable energy derived from cecal fermentation in the willow ptarmigan. Comp. Biochem. Physiol. A **53**(1), 115–121 (1976). https://doi.org/10.1016/s0300-9629(76)80022-9
10. Anwar, Z., Kausar, A., Rafique, I., Muhammad, B.: Advances in epoxy/graphene nanoplatelet composite with enhanced physical properties: a review. Polym.-Plast. Technol. Eng. **55**(6), 643–662 (2016). https://doi.org/10.1080/03602559.2015.1098695
11. Prolongo, S.G., Moriche, R., Jiménez-Suárez, A., Sánchez, M., Ureña, A.: Advantages and disadvantages of the addition of graphene nanoplatelets to epoxy resins. Eur. Polym. J. **61**, 206–214 (2014). https://doi.org/10.1016/j.eurpolymj.2014.09.022
12. Olszowska, K., et al.: Development of epoxy composites with graphene nanoplatelets and micro-sized carbon foam: morphology and thermal, mechanical and tribological properties. Tribol. Int. **185**, 108556 (2023). https://doi.org/10.1016/j.triboint.2023.108556
13. Tang, L.-C., et al.: The effect of graphene dispersion on the mechanical properties of graphene/epoxy composites. Carbon **60**, 16–27 (2013). https://doi.org/10.1016/j.carbon.2013.03.050
14. Chandrasekaran, S., Seidel, C., Schulte, K.: Preparation and characterization of graphite nano-platelet (GNP)/epoxy nano-composite: Mechanical, electrical and thermal properties. Eur. Polym. J. **49**(12), 3878–3888 (2013). https://doi.org/10.1016/j.eurpolymj.2013.10.008
15. Stankovich, S., et al.: Graphene-based composite materials. Nature **442**(7100), 282–286 (2006). https://doi.org/10.1038/nature04969
16. Balandin, A.A.: Thermal properties of graphene and nanostructured carbon materials. Nat. Mater. **10**(8), 569–581 (2011). https://doi.org/10.1038/nmat3064
17. Chrissafis, K., Bikiaris, D.: Can nanoparticles really enhance thermal stability of polymers? Part I: an overview on thermal decomposition of addition polymers. Thermochim. Acta **523**(1–2), 1–24 (2011). https://doi.org/10.1016/j.tca.2011.06.010
18. Potts, J.R., Dreyer, D.R., Bielawski, C.W., Ruoff, R.S.: Graphene-based polymer nanocomposites. Polymer **52**(1), 5–25 (2011). https://doi.org/10.1016/j.polymer.2010.11.042
19. Novoselov, K.S., Fal′ko, V.I., Colombo, L., Gellert, P.R., Schwab, M.G., Kim, K.: A roadmap for graphene. Nature **490**(7419), 192–200 (2012). https://doi.org/10.1038/nature11458
20. Coleman, J.N., et al.: Two-dimensional nanosheets produced by liquid exfoliation of layered materials. Science **331**(6017), 568–571 (2011). https://doi.org/10.1126/science.1194975
21. Mansor, M.R., Fadzullah, S.H.S.M., Masripan, N.A.B., Omar, G., Akop, M.Z.: Comparison between functionalized graphene and carbon nanotubes. In: Functionalized Graphene Nanocomposites and their Derivatives, pp. 177–204. Elsevier (2019). https://doi.org/10.1016/B978-0-12-814548-7.00009-X
22. Papageorgiou, D.G., Kinloch, I.A., Young, R.J.: Mechanical properties of graphene and graphene-based nanocomposites. Prog. Mater. Sci. **90**, 75–127 (2017). https://doi.org/10.1016/j.pmatsci.2017.07.004

23. Yu, Z., Wang, Z., Li, H., Teng, J., Xu, L.: Shape Memory Epoxy Polymer (SMEP) composite mechanical properties enhanced by introducing Graphene Oxide (GO) into the matrix. Materials **12**(7), 1107 (2019). https://doi.org/10.3390/ma12071107
24. The determination of the elastic field of an ellipsoidal inclusion, and related problems. Proc. Royal Soc. London. Ser. A. Math. Phys. Sci. **241**(1226), 376–396 (1997). https://doi.org/10.1098/rspa.1957.0133
25. Mori, T., Tanaka, K.: Average stress in matrix and average elastic energy of materials with misfitting inclusions. Acta Metall. **21**(5), 571–574 (1973). https://doi.org/10.1016/0001-6160(73)90064-3
26. Tornabene, F., Bacciocchi, M., Fantuzzi, N., Reddy, J.N.: Multiscale approach for three-phase CNT/polymer/fiber laminated nanocomposite structures. Polym. Compos. **40**(S1) (2019). https://doi.org/10.1002/pc.24520
27. Taş, H., Soykok, I.F.: Effects of carbon nanotube inclusion into the carbon fiber reinforced laminated composites on flexural stiffness: a numerical and theoretical study. Compos. Part B Eng. **159**, 44–52 (2019). https://doi.org/10.1016/j.compositesb.2018.09.055
28. El Bahi, A., et al.: Mechanical homogenization of transversely isotropic CNT/GNP reinforced biocomposite for wind turbine blades: numerical and analytical study. J. Compos. Sci. **7**(1), 29 (2023). https://doi.org/10.3390/jcs7010029
29. Ji, X.-Y., Cao, Y.-P., Feng, X.-Q.: Micromechanics prediction of the effective elastic moduli of graphene sheet-reinforced polymer nanocomposites. Model. Simul. Mater. Sci. Eng. **18**(4), 045005 (2010). https://doi.org/10.1088/0965-0393/18/4/045005

Plasmon Induced Enhanced Nonlinear Absorption in Vanadium Disulfide-Silver Nanocomposite

Hasana Jahan Elamkulavan[1], Sanjay Subramaniam[1], Jatinderbir Singh[1], Pacheri Madathil Aneesh[2], and Chandrasekharan Keloth[1](✉)

[1] Laser and Nonlinear Optics Laboratory, Department of Physics, National Institute of Technology, Calicut 673601, India
csk@nitc.ac.in

[2] Department of Physics, Central University of Kerala, Tejaswini Hills, Periye, Kasargod, Kerala 671320, India

Abstract. This study presents a novel vanadium disulfide-silver nanocomposite synthesized through liquid phase exfoliation, demonstrating a six-fold enhancement in nonlinear absorption coefficient compared to bare vanadium disulfide nanosheets. Characterization techniques confirm the formation of high-quality vanadium disulfide nanosheets, and UV-visible absorption spectra of the nanocomposite reveal significantly enhanced absorption in the range of 360 to 700 nm due to plasmon coupling and transfer of hot electrons. Open aperture Z-scan measurements demonstrate amplified reverse saturable absorption behavior in the nanocomposite which may be attributed to the transfer of hot electrons through plasmon coupling, leading to intensified field localization at the VS_2-Ag interface The observed enhancement underscores the efficacy of plasmon coupling in vanadium disulfide, offering potential applications in nonlinear optical devices.

Keywords: TMDs · VS_2-Ag nanocomposite · Plasmon coupling · Nonlinear Optics · Hybrid structures

1 Introduction

Nonlinear optics explores intense optical interactions occurring within bulk optical (3D) media. While they often face challenges like low carrier mobility and low recombination rates leading to power loss and complex nonlinear effects, 2D materials offer unique electronic and optical properties due to their atomically thin structure, resulting in reduced dielectric shielding and enhanced Coulomb interactions [1–3]. As highlighted in recent literature, the spatial confinement of charge carriers in 2D materials leads to quantized energy levels and discrete electronic states, resulting in distinctive quantum confinement effects [4]. Their atomically thin structures offer a high surface-to-volume ratio, enhancing light-matter interactions and enabling strong nonlinear effects even at low thicknesses. In addition to this, their thickness allows the 2D material to have highly

A. Tiwari et al. (Eds.): iConFIRM 2023, SPPHY 308, pp. 165–174, 2024.
https://doi.org/10.1007/978-981-97-4557-9_17

localized electric and magnetic fields, especially along edges and defects [5]. Capitalizing on the broad absorption spectrum and significant nonlinear optical response of 2D materials, they find extensive utility in various nonlinear applications including optical communication, frequency conversion, imaging, electro-optic modulation, mode-locking, and Q-switching [6–12]. Furthermore, their seamless integration with existing photonic and electronic platforms ensures compatibility with conventional fabrication processes, fostering the development of hybrid devices [13–15].

The emergence of Transition Metal Dichalcogenides (TMDs) such as MoS_2, $MoSe_2$, $MoTe_2$, WS_2, WSe_2, and TiS_2, presents a novel platform for investigating a variety of nonlinear optical effects [16–19]. They exhibit strong nonlinear optical responses, which are significant for applications in photonic and optoelectronic devices [12]. TMDs are 2D materials with exceptional mechanical and optical properties and the tailoring of these properties is nowadays, a key focus of research.

Reverse saturable absorption (RSA) is a nonlinear optical phenomenon where a material's absorption increases as the incident light intensity rises, contrasting with conventional saturable absorption (SA). RSA characteristic of the material aids a wide range of applications. It is common in places where modulation and control of light are necessary. One important application is optical limiting, where this material acts as a limiter to protect sensitive laser equipment. Recent research explores the superior optical limiting activity of titanium disulfide (TiS_2) nanosheets compared to benchmark materials like fullerene (C_{60}) and graphene, paving the way for highly efficient 2D-dichalcogenide-based optical limiters, particularly at low fluences [4]. The intensity-dependent absorption of an RSA material can be used to control its transmittance, making it an effective optical switch. Literature reports the usage of TMDs for such switching applications [20, 21]. The intensity-dependent absorption can also aid in pulse modulation. TMDs are found to be useful for such applications as well [17].

Vanadium disulfide (VS_2) is a versatile member of TMDs with semiconducting properties, making it adaptable for electronic devices and optoelectronics [22]. Its catalytic activity in hydrogen evolution reactions is valuable for catalysis, and its conductivity makes it suitable for energy storage applications [23, 24]. The material's potential in spintronics, and biomedical applications, and its intriguing electronic phase transitions highlight its importance across diverse technological domains, fueling ongoing research into its multifunctional properties [7, 25, 26]. While VS_2 has found applications in the optics field, notably in Q-switch pulsed lasers and mode-locking lasers, investigations into its nonlinear optical properties are in their early stages [27, 28].

Extensive research has been conducted on hybrid nanostructures comprising metal nanoparticles and 2D nanomaterials, aiming to enhance optical signals through plasmon amplification [29]. Plasmons signify the collective oscillations of conduction electrons in metals, exhibiting strongly confined fields around the nanoscale volume surface known as Localized Surface Plasmon Resonance (LSPR), which occurs at a specific frequency depending on the size, shape, and local environment of the metal nanostructure [5, 30].

A VS_2-silver nanocomposite (VS_2-Ag NC) is expected to possess distinctive properties arising from the combination of VS_2 and Ag nanoparticles. VS_2 is a semiconductor with promising electrical properties, while Ag nanoparticles exhibit plasmonic effects and catalytic activity. By incorporating Ag nanoparticles into the VS_2 matrix, the composite may exhibit enhanced electrical conductivity due to Ag's metallic nature. Additionally, the presence of Ag nanoparticles could induce plasmonic effects, potentially enhancing optical properties such as absorption and scattering. The interaction between VS_2 and Ag nanoparticles may also lead to enhanced structural stability, providing mechanical robustness and resistance to degradation. The tunable nature of the composite, achievable by adjusting parameters such as nanoparticle size, shape, and concentration, may offer opportunities for tailoring its properties for specific applications in electronics, catalysis, sensing, and nonlinear optics.

In this research, the versatile properties of hydrothermally grown VS_2 nanosheets and chemically synthesized plasmonic silver nanoparticles are combined using a simple liquid phase exfoliation method to introduce a novel vanadium disulfide-silver (VS_2-Ag) nanocomposite (NC) having distinguishably enhanced nonlinear absorption (NLA) properties.

2 Materials and Method

2.1 Preparation of the Samples

VS_2 powder was synthesized using a simple hydrothermal method reported in our previous work [27]. In short, 8 mM ammonium metavanadate (NH_4VO_3) was dissolved in a solution containing 50 mL of distilled water and 12 mL of ammonia. 64 mM thioacetamide (C_2H_5NS) was separately stirred in 10 mL of distilled water. The two solutions were thoroughly mixed and then kept for hydrothermal treatment for 24 h at a temperature of 180 °C in a hot air oven. After cooling to room temperature, the resulting black products were washed multiple times with distilled water and dried in an oven at 200 °C for 1 h.

To make the VS_2 nanosheets in water, 2.28 mg of the synthesized VS_2 powder was ultrasonicated in 8 ml of DI water using an ultrasonicator (Power Sonic 605) for 8 h at room temperature. The nanosheets appeared to be greenish. Ag nanoparticles are prepared chemically in water [31]. A solution is prepared by adding 5 mM trisodium citrate and 2.5 mM tannic acid in 50 ml of deionized (DI) water. Another solution is prepared by taking 25 mM $AgNO_3$ in 0.5 ml of DI water. The second solution is added dropwise to the first solution and stirred for 1 h to get a uniform Ag colloidal solution which appeared brownish.

To make the VS_2-Ag NC, 2.28 mg of VS_2 powder was ultrasonicated in 8 ml of the as-prepared Ag colloidal solution for 8 h at room temperature (Fig. 1a). To compare the optical properties, we have prepared the VS_2 nanosheets in water at the similar conditions of synthesis of VS_2-Ag NC. The nanocomposite was dark brownish-green in color, whereas pure VS_2 nanosheets appeared greenish. The photographs of synthesized VS_2 nanosheets in water, colloidal silver nanoparticles, and VS_2-Ag nanocomposite are given in Fig. 1b.

2.2 Characterization

To capture the X-ray diffraction pattern, a Rigaku Miniflex 600 diffractometer equipped with Cu Kα radiation ($\lambda = 1.5406$ Å) was used. Using a FE-SEM (ΣIGMA, ZIESS), the surface morphology of the nanostructures was examined. The materials' linear optical absorption properties were examined using a 'Shimadzu UV 2450' UV-visible spectrophotometer. Open aperture Z-scan technique was used to investigate the NLA properties (Fig. 1c). The laser we used was a Q-switched Nd: YAG laser of 532 nm wavelength, 7 ns pulse width, and a repetition rate of 10 Hz.

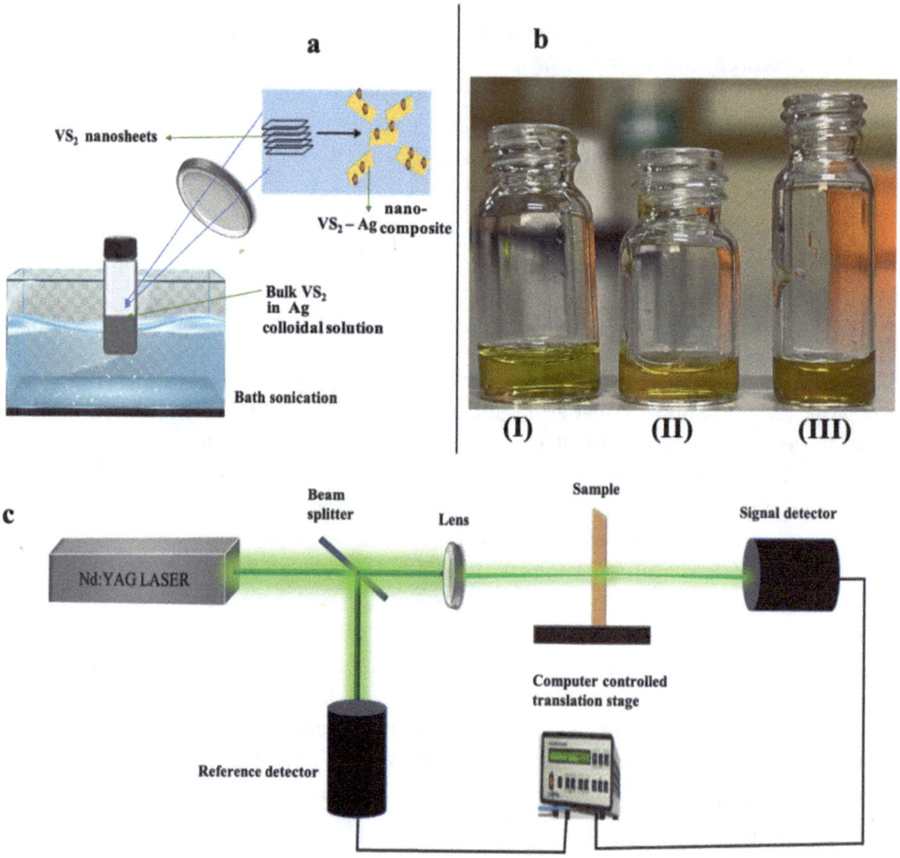

Fig. 1. Schematic diagrams of a) liquid phase exfoliation method for the synthesis of the NC b) photograph of (I) VS_2 nanosheets in water, (II) colloidal silver nanoparticles, and (III) VS_2-Ag nanocomposite and c) Z-scan setup

3 Results and Discussion

The X-ray diffraction pattern analysis was employed to confirm the formation of pure VS_2 nanostructures. Figure 2a displays the XRD pattern of the synthesized VS_2 nanostructures where the diffraction peaks align consistently with the hexagonal phase of VS_2 (JCPDS 01-089-1640). FE-SEM image shows the formation of a good-quality nanosheet-like structure of synthesized VS_2 (Fig. 2a). The thickness of the layers was calculated to be in the range of 25–30 nm.

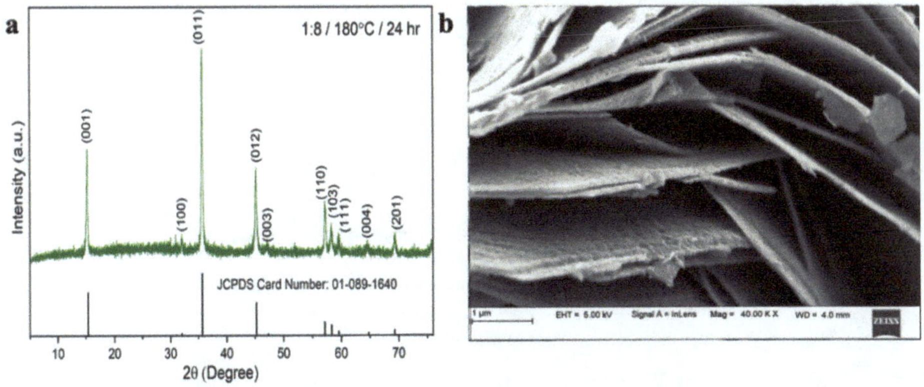

Fig. 2. a) XRD pattern of VS_2 nanostructures, and b) FESEM image of VS_2 nanostructures

Linear optical responses of the VS_2 nanosheets in water, colloidal silver nanoparticles, and VS_2-Ag nanocomposite are recorded to confirm the formation of the nanocomposite. As shown in Fig. 3, pure VS_2 nanosheets in water exhibit an absorption peak around 385 nm wavelength, which may be due to excitonic transitions [27]. However, it has almost negligible absorption in the range of 500 to 700 nm. The absorption spectrum of Ag nanoparticles exhibits a plasmonic band in the range of 360–470 nm (LSPR peak at 407 nm) [32]. LSPR refers to the collective oscillation of conduction band electrons in the metal when excited by incident light [29]. In the case of silver nanoparticles, LSPR typically occurs in the visible region of the electromagnetic spectrum due to their unique size, shape, and composition-dependent optical properties. When the NC is formed, the LSPR of Ag overlaps with the excitonic transition of VS_2 nanosheets. It is evident from comparing the graphs that, the absorption spectrum showed a broader and significantly enhanced absorption in the range of 360 to 700 nm for VS_2-Ag nanocomposite, which may be due to the collective SPR absorption of the Ag nanoparticles and excitonic transition of VS_2 nanosheets. The dielectric functions of Ag and VS_2 are reported in the literature [33, 34].

A standard open aperture Z-scan technique, developed by Sheikh Bahae [35], was used to investigate the nonlinear absorption behavior of the samples. The method involves moving the sample along the z-axis while exposing it to a tightly focused Gaussian beam of radius w_0, with $z_0 = \frac{\pi w_0^2}{\lambda}$ being the Rayleigh range. The nonlinear pulse propagation equation can be integrated and used for a numerical fit of the transmitted intensity about the position z

$$\frac{dI}{dz'} = -\left[\frac{\alpha_0}{1+\frac{I}{I_s}} + \beta_{eff} I\right] I \tag{1}$$

The transmittance curve depends on two variables, namely the saturation intensity (I_s) and the effective nonlinear absorption coefficient (β_{eff}). These variables are related

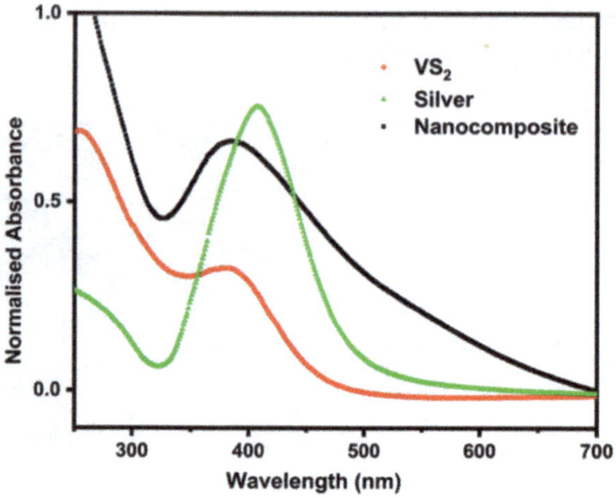

Fig. 3. A comparison of UV-visible absorption spectra of silver nanoparticles, VS_2 nanosheets, and VS_2-Ag nanocomposites

to the linear absorption coefficient (α_0), incident intensity (I), and the sample propagation distance (z′). I_s primarily affects the transmittance values above one, while β_{eff} is responsible for the transmittance dip.

Figure 4 illustrates the nonlinear optical responses of the synthesized structures at 0.28 GW/cm^2 on-axis intensity. When the samples are positioned far from the focal point, they demonstrate no nonlinearity, and their transmittance remains at unity under weak intensity. However, as the samples are brought closer to the focus, the nonlinear effects come into play. In Ag nanoparticles, the transmittance increases near the focus, leading to the SA behavior [36]. On the other hand, the transmittance of pure VS_2 nanosheets shows the transmittance dip as it approaches the focal point (z = 0 mm) signifying a reverse saturable absorption. At the same on-axis intensity, the VS_2-Ag nanocomposite displays amplified nonlinear absorption behavior. This is evidenced by a sharp decrease in transmittance, signifying the presence of exceptionally strong RSA behavior compared to the bare VS_2 nanosheets. Comparing the NLA coefficients for the samples given in Table 1, we can summarize that the observed NLA coefficient of the nanocomposite is 6.25 times greater than that of bare VS_2 nanosheets. Therefore, we obtained more than 6-fold enhancement in the optical nonlinearity by solely making the nanocomposite of VS_2. Table 1 shows a comparison of nonlinear absorption coefficients of similar structures reported in the literature. It indicates that our nanocomposite is showing nonlinear optical (NLO) properties better or on par with similar reported structures.

The enhanced nonlinear optical behavior observed in the VS_2-Ag nanocomposite can be understood by observing the UV-visible absorption spectra of the samples and may be attributed to the significant absorption of VS_2-Ag NC at the wavelength of excitation (532 nm) and the transfer of hot electrons. Under illumination at 532 nm, the incident photons are intensely confined and amplified at interfacial plasmonic hot spots due to plasmon coupling with VS_2 nanosheets [41]. The excited localized surface plasmon resonances undergo rapid decay via Landau damping, leading to the creation

Table 1. A Comparison of NLO Parameters of the VS_2, Ag, and VS_2-Ag nanocomposite with similar structures

Sample	Laser Wavelength (nm)	β_{eff}(cm/GW)
Pure VS_2 nanosheets †	532	20
Silver †	532	–
VS2-Ag NC †	532	125
TiS_2 sheets [18]	532	80
GO [37]	532	20
C_{60}[38]	532	40
$MoSe_2$/GO Hybrid [37]	532	580
MWNT-OH [38]	532	52
MWNT-COOH [38]	532	57
WS_2/PMMA [39]	532	44.46
h-BN [40]	1064	74.84
Graphene sheets [18]	532	30.1
MWNT [38]	532	55

† This work.

of hot electrons. These electrons can cross the Schottky barrier height and be transferred into the unoccupied conduction band of VS_2, while concurrently leaving a hole in the silver. The involvement of these high-energy hot electrons is pivotal in enhancing the Excited State Absorption (ESA) in VS_2 nanosheets and fortifying the RSA behavior of the nanocomposites [42–45].

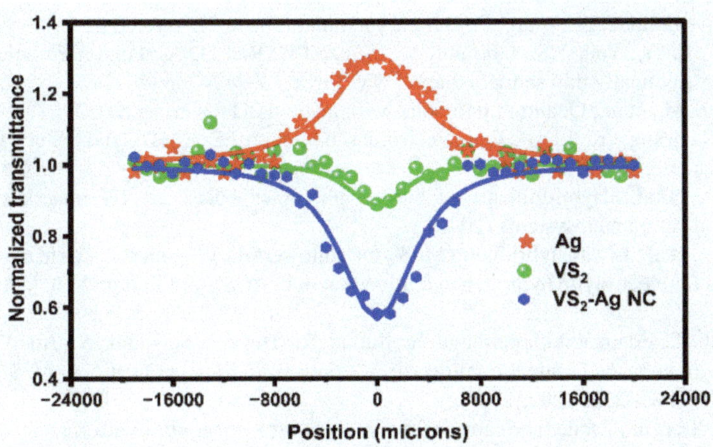

Fig. 4. A comparison of the open aperture Z-scan curves for the samples

4 Conclusion

In summary, this work successfully synthesized and characterized a Vanadium disulfide-silver nanocomposite with enhanced nonlinear absorption properties. The combination of hydrothermally grown VS_2 nanosheets and chemically synthesized plasmonic silver nanoparticles resulted in a nanocomposite exhibiting significantly amplified reverse saturable absorption behavior. The ESA-assisted six-fold enhancement in the nonlinear optical coefficient highlights the effectiveness of plasmon coupling. These findings contribute to the potential applications of VS_2-Ag nanocomposites in nonlinear optical devices, emphasizing their role in advancing composite nanostructures for enhanced optical functionalities.

References

1. Cipriano, L.A., Di Liberto, G., Tosoni, S., Pacchioni, G.: Quantum confinement in group III-V semiconductor 2D nanostructures. Nanoscale **12**, 17494–17501 (2020)
2. Gupta, A., Sakthivel, T., Seal, S.: Recent development in 2D materials beyond graphene. Prog. Mater Sci. **73**, 44–126 (2015)
3. Tareen, A.K., Khan, K., Aslam, M., Liu, X., Zhang, H.: Confinement in two-dimensional materials: major advances and challenges in the emerging renewable energy conversion and other applications. Progress Solid State Chem. **61** (2021)
4. Varma, S.J., et al.: 2D TiS_2 layers: a superior nonlinear optical limiting material. Adv. Opt. Mater. **5** (2017)
5. Fan, X., Zheng, W., Singh, D.J.: Light scattering and surface plasmons on small spherical particles. Light: Sci. Appl. **3** (2014)
6. Li, L., Pang, L., Zhao, Q., Wang, Y., Liu, W.: VS_2 as saturable absorber for Q-switched pulse generation. Nanophotonics. **9**, 2569–2576 (2020)
7. Niu, Z., et al.: Layered metallic vanadium disulfide for doubly Q-switched Tm:YAP laser with EOM: experimental and theoretical investigations. Nanomaterials **11** (2021)
8. Guo, B., et al.: Dual-Wavelength soliton mode-locked fiber laser with a WS_2-based fiber taper. IEEE Photonics Technol. Lett. **28**, 323–326 (2016)
9. Sun, Z., Martinez, A., Wang, F.: Optical modulators with 2D layered materials (2016)
10. Liu, H., Li, Y., You, Y.S., Ghimire, S., Heinz, T.F., Reis, D.A.: High-harmonic generation from an atomically thin semiconductor. Nat. Phys. **13**, 262–265 (2017)
11. Turunen, M., et al.: Quantum photonics with layered 2D materials (2022)
12. Huo, N., Yang, Y., Li, J.: Optoelectronics based on 2D TMDs and heterostructures. J. Semiconduct. **38** (2017)
13. You, J., et al.: Hybrid/integrated silicon photonics based on 2D materials in optical communication nanosystems (2020)
14. Yu, S.H., et al.: In Situ hybridizing MoS_2 microflowers on VS_2 microflakes in a one-pot CVD process for electrolytic hydrogen evolution reaction. ACS Appl Energy Mater. **2**, 5799–5808 (2019)
15. Abhijith, T., Edappadikkunnummal, S., Suthar, R., Thomas, S., Karak, S.: Au-WS_2 nanohybrids with enhanced optical nonlinearity for optical limiting applications. ACS Appl. Nano Mater. **6**, 2327–2335 (2022)
16. Zhang, X., et al.: Facile fabrication of wafer-scale MoS_2 neat films with enhanced third-order nonlinear optical performance. Nanoscale **7**, 2978–2986 (2015)
17. Li, T., et al.: Third-order nonlinear effect and polarization-dependent modulation using spherical-like MoS_2 nanoparticles. Ann. Phys. **535** (2023)

18. Hernandez-Rueda, J., Noordam, M.L., Komen, I., Kuipers, L.: Nonlinear optical response of a WS_2 monolayer at room temperature upon multicolor laser excitation. ACS Photon. **8**, 550–556 (2021)
19. Lu, C., et al.: Layer-dependent nonlinear optical properties of WS_2, MoS_2, and Bi_2S_3 films synthesized by chemical vapor deposition. ACS Appl. Mater. Interfaces **14**, 2390–2400 (2022)
20. Wu, K., Zhang, X., Wang, J., Li, X., Chen, J.: WS_2 as a saturable absorber for ultrafast photonic applications of mode-locked and Q-switched lasers. Opt. Express **23**, 11453–11461 (2015)
21. Klein, J., et al.: Electric-field switchable second-harmonic generation in bilayer MoS2 by inversion symmetry breaking. Nano Lett. **17**, 392–398 (2017)
22. Sahoo, D., Senapati, S., Naik, R.: Progress and prospects of 2D VS_2 transition metal dichalcogenides (2022)
23. Kumar, G.M., Ilanchezhiyan, P., Cho, H.D., Lee, D.J., Kim, D.Y., Kang, T.W.: Ultrathin VS2 nanodiscs for highly stable electro catalytic hydrogen evolution reaction. Int. J. Energy Res. **44**, 811–820 (2020)
24. Zhang, X., et al.: Insights into the storage mechanism of layered vs2 cathode in alkali metal-ion batteries. Adv. Energy Mater. **10** (2020)
25. Yue, G., Liu, X., Mao, Y., Zheng, H., Zhang, W.: A promising hybrid counter electrode of vanadium sulfide decorated with carbon nanotubes for efficient dye-sensitized solar cells. Mater. Today Energy. **4**, 58–65 (2017)
26. Littlejohn, A.J., et al.: Large metallic vanadium disulfide ultrathin flakes for spintronic circuits and quantum computing devices. ACS Appl. Nano Mater. **2**, 3684–3694 (2019)
27. Li, L., Pang, L., Zhao, Q., Wang, Y., Liu, W.: VS2as saturable absorber for Q-switched pulse generation. Nanophotonics **9**, 2569–2576 (2020)
28. Joseph, A., Elamkulavan, H.J., Keloth, C., Aneesh, P.M.: Hydrothermally grown VS_2 nanosheets: a material for optical limiting applications. ACS Appl. Opt. Mater. **1**, 1688–1696 (2023)
29. Anand, B., Kaniyoor, A., Sai, S.S.S., Philip, R., Ramaprabhu, S.: Enhanced optical limiting in functionalized hydrogen exfoliated graphene and its metal hybrids. J. Mater. Chem. C **5** (2013)
30. Amendola, V., Pilot, R., Frasconi, M., Maragò, O.M., Iatì, M.A.: Surface plasmon resonance in gold nanoparticles: a review (2017)
31. Ranoszek-Soliwoda, K., et al.: The role of tannic acid and sodium citrate in the synthesis of silver nanoparticles. J. Nanoparticle Res. **19** (2017)
32. Hari, M., Mathew, S., Nithyaja, B., Joseph, S.A., Nampoori, V.P.N., Radhakrishnan, P.: Saturable and reverse saturable absorption in aqueous silver nanoparticles at off-resonant wavelength. Opt. Quant. Electron. **43**, 49–58 (2012)
33. The optical properties of bulk and (0 0 1) monolayer VS_2 based on first principles calculations. Acad. J. Mater. Chem. **3** (2022)
34. Johnson, P.B., Christy, R.W.: Optical constants of the noble metals. Phys. Rev. B **6**, 4370–4379 (1972)
35. Sheik-Bahae, M., Said, A.A., Wei, T.-H., Hagan, D.J., Van Stryland, E.W.: Sensitive measurement of optical nonlinearities using a single beam (1990)
36. Jiang, J., Zuo, S., Wang, G., Liu, J., Zhang, X., Gao, Y.: Double transformation of the nonlinear absorption in silver nanoparticles. Opt. Express **30**, 41255 (2022)
37. Sharma, R., et al.: Strong interlayer coupling mediated giant two-photon absorption in $MoSe_2$/graphene oxide heterostructure: quenching of exciton bands. Phys. Rev. B. **93** (2016)
38. Zhang, X.L., Liu, Z.B., Yan, X.Q., Li, X.C., Chen, Y.S., Tian, J.G.: Nonlinear optical and optical limiting properties of fullerene, multi-walled carbon nanotubes, graphene and their derivatives with oxygen-containing functional groups. J. Opt. (United Kingdom). **17** (2015)

39. Long, H., et al.: Tuning nonlinear optical absorption properties of WS_2 nanosheets. Nanoscale **7**, 17771–17777 (2015)
40. You, J.W., Bongu, S.R., Bao, Q., Panoiu, N.C.: Nonlinear optical properties and applications of 2D materials: theoretical and experimental aspects (2018)
41. Jia, C., et al.: Interface-engineered plasmonics in metal/semiconductor heterostructures. Adv. Energy Mater. **6** (2016)
42. Wen, X., Xu, W., Zhao, W., Khurgin, J.B., Xiong, Q.: Plasmonic hot carriers-controlled second harmonic generation in WSe_2 Bilayers. Nano Lett. **18**, 1686–1692 (2018)
43. Park, Y., et al.: Elongated lifetime and enhanced flux of hot electrons on a perovskite plasmonic nanodiode. Nano Lett. **19**, 5489–5495 (2019)
44. Cushing, S.K., et al.: Controlling plasmon-induced resonance energy transfer and hot electron injection processes in Metal@TiO_2 core-shell nanoparticles. J. Phys. Chem. C **119**, 16239–16244 (2015)
45. Yu, Y., et al.: Ultrafast plasmonic hot electron transfer in au nanoantenna/MoS_2 heterostructures. Adv. Funct. Mater. **26**, 6394–6401 (2016)

Synthesis and Characterization of Chitosan Functionalized Graphene-Oxide Based Binary Metal Composite for Energy Storage Application

Gaytri, Pankaj Kumar, and Partha Khanra(✉)

Institute of Engineering and Technology, Chitkara University, Patiala 140401, Punjab, India
partha.khanra@chitkara.edu.in

Abstract. Chitosan functionalized graphene oxide (GO) based binary metal composites were synthesized via the hydrothermal approach for energy storage applications. The composite material demonstrated a notable increase in surface area by successful cross-linking between the amine groups present in chitosan and the carboxylic groups in GO, and a significant dispersion of metal nanoparticles on the surface of GO was observed in the presence of chitosan. Interestingly, in the FTIR spectra, two dominating peaks appeared at 1388 cm^{-1} and 1161 cm^{-1} which attributed to the absorption peak of the glycosidic bond, and stretching vibrations of C–O bending respectively. The structural analysis was confirmed by XRD whereas, the broadened peak at 26° indicated the (002) plane of reduced graphene which is similar peak of typical graphite. The presence of metal nanoparticles and chitosan generates the higher separation of two graphitic planes. Moreover, the indexed (111), (220), (311), (400), (422), (511), (440), and (533) planes indicate the spinel structure of $NiCo_2O_4$. Noticeably, the electrochemical performances were carried out by analyzing potentiometric, galvanometric, and electrochemical impedance. In this study, we have observed that the potentiometric curves appeared with two distinct redox peaks which indicates the electrochemical redox reactions between metal compounds and aqueous electrolytes.

Keywords: Supercapacitor · Graphene Oxide · Binary Metal Composite

1 Introduction

When it comes to meeting the ever-increasing demands of energy storage devices, portable electronics, and electric cars, the development of innovative energy materials is necessary [1]. Supercapacitors, known for their remarkable energy and power density, long lifespan, environmental friendliness, and safety features, are a leading choice in power storage solutions. These devices can be classified into three types, viz., electrochemical double-layer capacitors (EDLCs), pseudo-capacitors, and hybrid supercapacitors, each offering diverse charge storage mechanisms. In general, the capacitor consists of three basic parts, whereas the scientific community coincides with their concentration for the betterment of energy storage performances, such as (i) electrode; (ii) electrolyte;

A. Tiwari et al. (Eds.): iConFIRM 2023, SPPHY 308, pp. 175–182, 2024.
https://doi.org/10.1007/978-981-97-4557-9_18

and (iii) separator. In this report, the development of the electrode materials are highlighted. Various types of electrode materials are reported, which can be categorized into three different types, for an example, (i) carbon-based electrode, which is highly effective for EDCLs; (ii) transition metal oxide or sulfide-based which is effective for pseudo capacitor; and (iii) carbon-metal based composites which is effective for hybrid capacitor. In this study, carbon materials such as graphene-based metal composite have been prepared to study the electrochemical performances. Because, graphene oxide (GO), a remarkable carbonaceous material, has garnered attention in supercapacitor development due to its high-enhanced conductivity, high dispersibility, and chemical stability [2–5]. Despite these advantages, the potential of GO as an energy storage material is hindered by the stacking of graphene sheets, leading to reduced porosity [6].

To overcome this limitation, we have introduced chitosan, a naturally found polysaccharide with diverse surface functional groups, including amine and hydroxyl groups. Here, a sandwich-like arrangement of chitosan/graphene oxide complex (CSGO) was prepared. Notably, chitosan was employed without any treatment and subsequently doped with nickel-cobalt salts using the hydrothermal technique. Potassium hydroxide (KOH) was used as the carrier electrolyte, whereas, potassium ferrocyanide ($K_4[Fe(CN)_6]$) was used as a redox-active electrolyte. This novel CSGO complex is designed to enhance the overall performance of supercapacitors by addressing the porosity issue associated with the stacking of graphene sheets. Through these efforts, we aim to contribute to the advancement of energy storage materials, improving their efficiency and applicability in various applications.

2 Materials and Methods

2.1 Materials

Graphene oxide (GO) powder with 99% purity was procured from Platonic Nano-tech Private Limited, Mohanpur Dist. Godda, Jharkhand, India. Chitosan and polyvinylidene fluoride (PVDF) were obtained from Sigma Aldrich Co. (Japan). Nickel chloride ($NiCl_2{\cdot}6H_2O$, 99%, extrapure, AR), cobalt chloride ($CoCl_2{\cdot}6H_2O$, 99%, AR), acetic acid (99.9% extrapure, AR) were purchased from Sisco Research Laboratories Pvt. Ltd., New Mumbai, India. Potassium hydroxide (KOH, 85%, AR), N-methyl-2-pyrrolidone (NMP, 98%), and ethanol (EtOH) were acquired from Loba Chemie Pvt. Ltd., Mumbai, India. Nickel foam ($\geq$ 99.6%) and conducting carbon black (CB) were purchased from Changsha Lyrun Material Co., Ltd., China.

2.2 Synthesis of Chitosan-Modified GO Binary Metal Composite (CSGO-M)

The synthesis of chitosan-modified GO with nickel-cobalt oxide composite was successfully achieved using a straightforward hydrothermal technique (Fig. 1). In the present experiment the GO (25 mg) was uniformly dispersed in 50 ml of water using ultrasonication. Subsequently, 10 ml chitosan (1 mg/ml) aqueous solution was prepared by the addition of 1 ml acetic acid. The prepared chitosan solution was carefully added to the as-prepared GO solution under the sonication environment. Then, 5 wt% each of

$NiCl_2.6H_2O$ and $CoCl_2.6H_2O$ salts were added into the solution and kept under continuous stirring. Following a period of uninterrupted agitation, the solution containing dispersed particles was transferred to a Teflon-lined vessel subjected to a temperature of 180 °C for 12 h. The resulting black material was centrifuged and washed multiple times with an equimolar solution of ethanol and water. The washed product was then dried overnight under a vacuum before being stored in a dry cabinet for further use. For reliable comparison, pristine GO metal binary composite (GO-M), and chitosan-modified GO (CSGO) were prepared using the above methodology in the absence of chitosan and metal salts respectively.

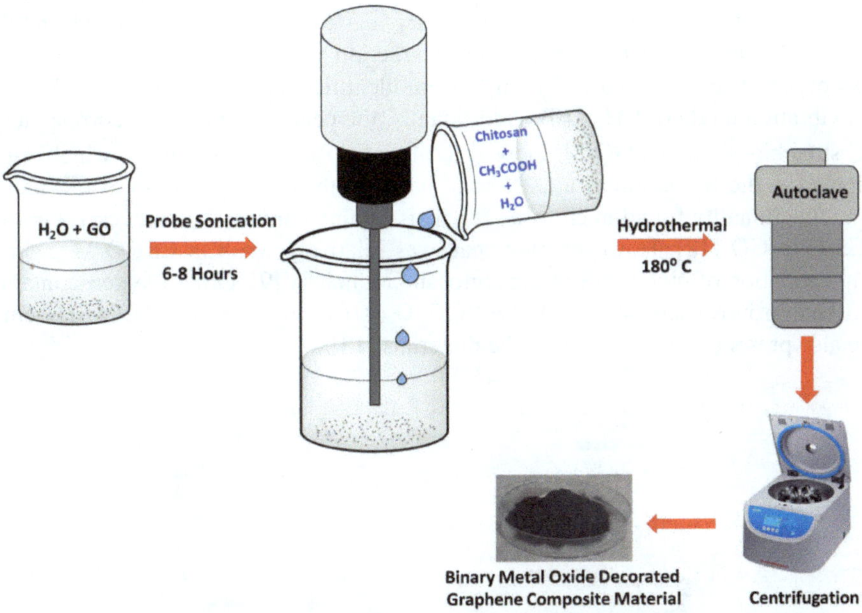

Fig. 1. Synthesis of chitosan-modified GO binary metal composite (CSGO-M) using hydrothermal method

2.3 Electrochemical Analysis

For the electrochemical study 2 M KOH aqueous electrolyte was used and later on for the improvement of the capacitance performance 0.1 M $K_4[Fe (CN)]_6$ was added with it. The electrochemical setup was purchased from Won Tech., South Korea (Model Zive Sp1), whereas cyclic voltammetry (CV), and charge-discharge (GCD) studies were carried out. The operating potential window was –0.2 V to 0.48 V for the CV study and the scan rates ranged from 10 to 100 mVs^{-1}. The GCD measurements encompassed different current densities, specifically 1 to 10 A g^{-1}, within the potential range of 0 to 0.40 V. This systematic electrochemical investigation aimed to comprehensively characterize the materials under diverse electrolytic conditions, providing valuable insights into their performance and behavior across varying scan rates and current densities [7].

3 Results and Discussion

3.1 Characterization of Chitosan-Modified GO Binary Metal Composite

The chitosan-modified GO binary metal composite (CSGO-M) was successfully synthesized by a one-step hydrothermal technique. For comparison study, GO-M was synthesized to elucidate the importance of chitosan modification on the electrochemical performance of the as-synthesized binary metal composite. The surface properties and chemical structure of the materials were studied via FTIR spectroscopy (Fig. 2)

A wide absorption band in the range of 3200 - 3600 cm^{-1} was observed in all three samples suggesting the involvement of O–H or/and N–H stretching [8, 9]. Specifically, the intensity of the band decreases for binary metal composites (CSGO-M and GO-M), possibly due to the hydroxyl group interaction with nickel and cobalt ions. The peaks observed at 2924 and 2874 cm^{-1} were identified as the adsorption peaks of the C–H vibration mode of CH_2 [10]. Additionally, the peak at 1560 cm^{-1} corresponds to C=C stretching [11]. CSGO-M showed an intense peak at 1388 cm^{-1}, which can be attributed to the formation of a C–N bond between the carboxyl group of GO and the amine functionality found in chitosan [8]. This finding confirms the successful grafting of CS onto GO [12]. However, this peak was relatively weak in CSGO-M owing to the intercalation of metal ions in the chitosan-GO matrix [9]. Other oxygen-containing functional groups such as C = O (1700), C–O (1161 cm^{-1}), and C–OH (1020 cm^{-1}) were also present on the surface of the materials [13].

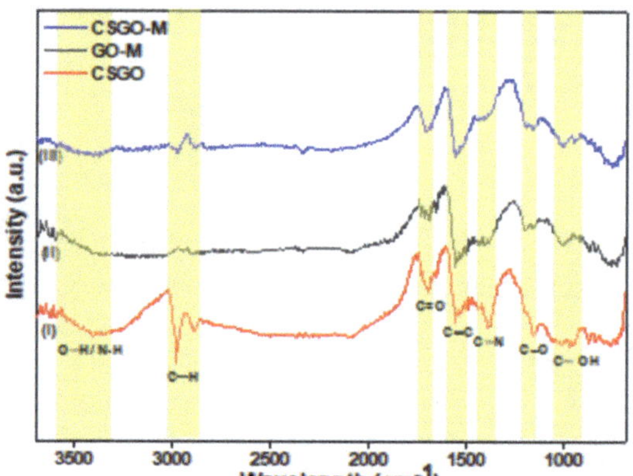

Fig. 2. FTIR spectra of (i) CSGO, (ii) GO-M, and (iii) CSGO-M

Both GO-binary metal composites and their chitosan-modified equivalents were subjected to X-ray diffraction (XRD) analysis to investigate the crystallographic properties of the composites. The characteristic peak for GO at 12° was found to be shifted to 26°, indicating the reduction of oxygen functionalities on the surface of GO to form graphene layers. The (002) peak in the case of CSGO-M was more pronounced as compared to GO-M, this could be attributed to further reduction of GO in the presence of chitosan by forming the amide linkages [14].

Additionally, a few other diffraction peaks corresponding to, (111), (220), (311), (400), (422), (511), (440), and (533) planes with low intensity suggested the presence of spinel $NiCo_2O_4$ (JCPDF No. 73–1702) on the surface of GO [15]. The Raman spectra of CSGO-M and GO-M revealed two prominent bands at 1342 cm^{-1} (D band) and 1580 cm^{-1} (G band), suggesting the existence of a graphene structure in the binary metal composites. Furthermore, there are additional minor peaks observed in the 400 to 700 cm^{-1} range (as shown in the inset of Fig. 3 (a)), which correspond to specific vibrational modes of spinel $NiCo_2O_4$ [16].

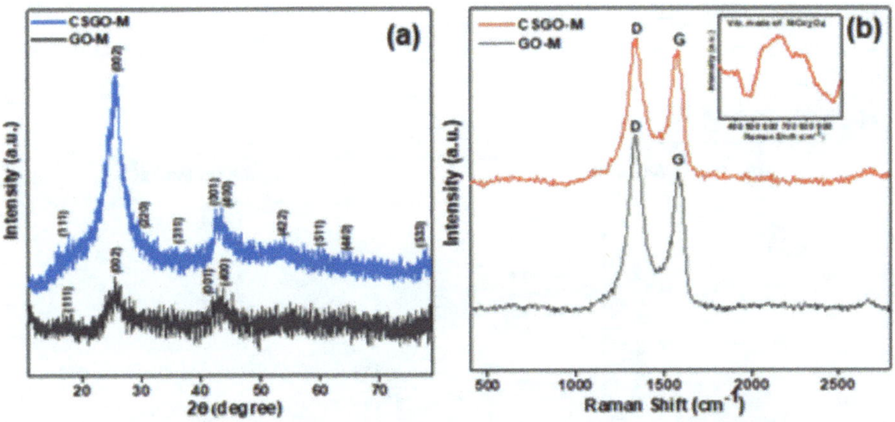

Fig. 3. (a) XRD spectra and (b) Raman spectra of GO-M and CSGO-M

3.2 Electrochemical Performance of Chitosan-Modified GO Binary Metal Composite

The composite's electrochemical performance was assessed by CV and GCD (Figs. 4, 5). CV measurements were conducted for both materials at different scan rates in the range 5 to 100 mV s^{-1}) in electrolytes containing KOH and KOH + $K_4[Fe(CN)_6]$. The redox peaks were observed for both samples in 1M KOH electrolyte solution. Specifically, for CSGO-M, the oxidative peak and redox peak were observed at 0.44 V and 0.27 V respectively. Overall, the CV area curve appeared semi-rectangle and the CV area increased significantly when the scan rates were increased, which was subsequently accompanied by a progressive increase in the current density. A similar trend in the CV curve area was observed when current density was increased. Interestingly, the oxidation and reduction peaks did not shift drastically with an increase in scan rates indicating that the redox reactions were reversible. The electron transition from the redox center to the electrode is advantageous for supercapacitor application. At high scan rates, the redox-active compounds exhibited insufficient diffusion time, resulting in a decrease in specific capacitance. However, when it came to the chitosan-modified binary metal composite (CSGO-M), the impact was not as significant, indicating a quicker transfer of charge and reduced resistance upon contact in this scenario [17]. This comparative study highlighted the increase in inter-layer spacing of GO-based binary metal composite thus ensuring the uniform dispersion of binary metal particles.

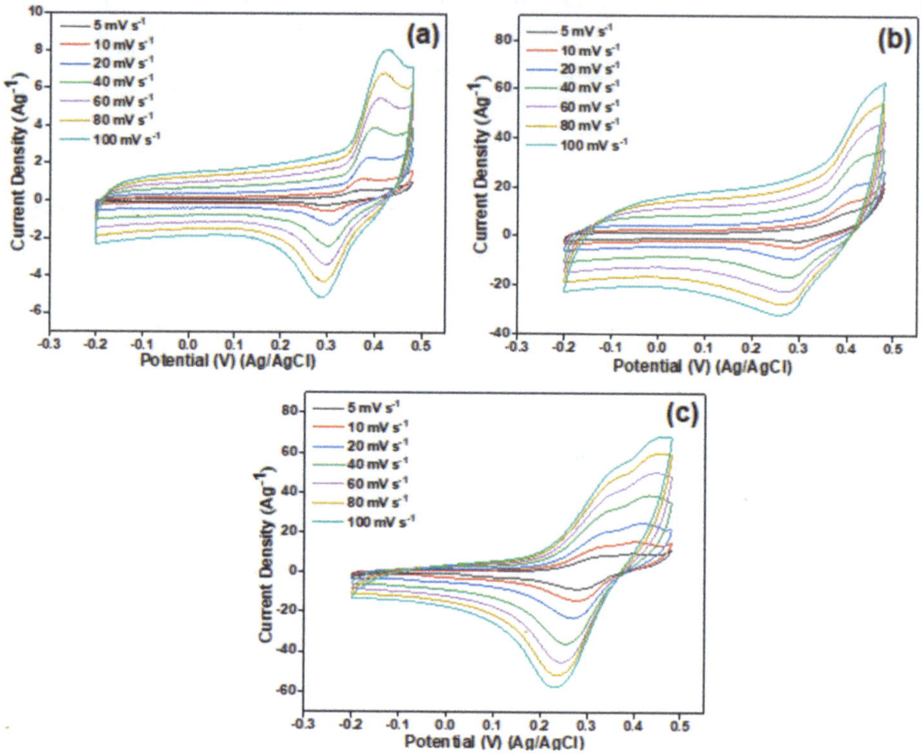

Fig. 4. CV curves of (a) GO-M-KOH, (b) CSGO-M-KOH, and (c) CSGO-M-KOH + $K_4[Fe(CN)_6]$ at different scanning rates 5, 10, 20, 40, 60, 80 and 100 mV s^{-1}

In comparison to CSGO, CSGO-M displayed a much bigger enclosed CV area, which is indicative of a greater specific capacitance. There was a significant increase in the specific capacitance of CSGO-M in the presence of KOH + $K_4[Fe(CN)_6]$ to 777.5 F g^{-1} when the current density was 1.5 A g^{-1} (Fig. 5 (c, d)). This contrasts with the fact that the specific capacitance in KOH was only 200 F g^{-1}. With the addition of potassium ferrocyanide to the electrolyte, the specific capacitance of CSGO-M was significantly increased. In the presence of potassium ferrocyanide, CSGO-M showed two distinct peaks at 0.35V and 0.44V however, the reduction peaks were merged at 0.23 V.

This is attributed to the increased diffusion of Fe (II)/Fe (III)-redox couple. The GCD curves (Fig. 5 (b)) of all the materials showed apparent plateaus in a distorted triangle-like structure during the charging/discharging process. The curves show a symmetric nature at a current density of 2 A g^{-1}. Further, it was noted that the discharge time increased in the case of chitosan-modified graphene oxide metal composite (CSGO-M), concerning GO-M.

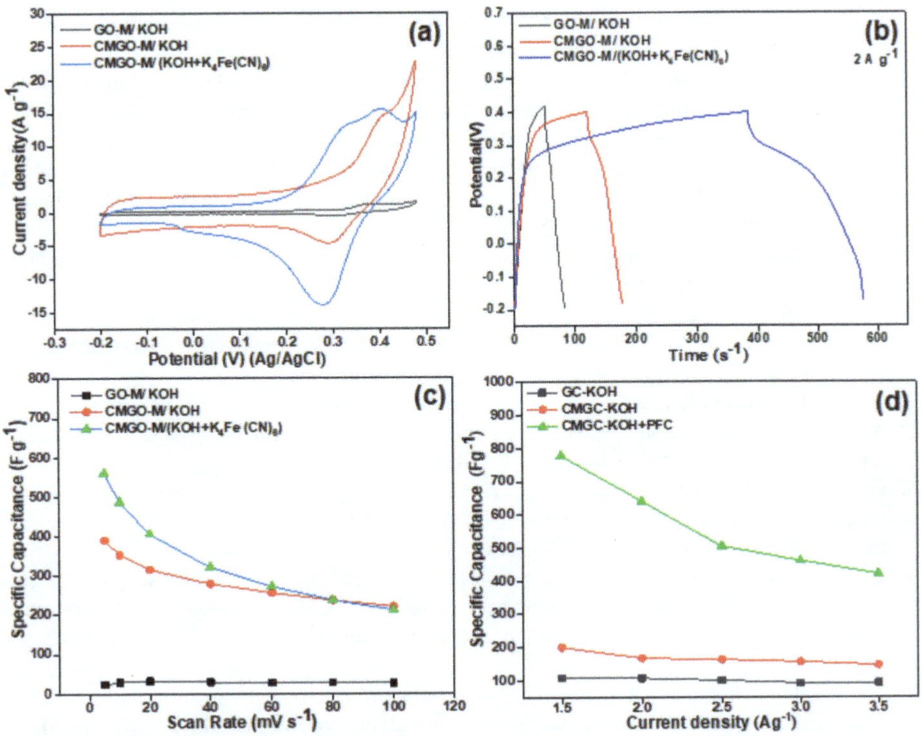

Fig. 5. (a) CV curves at a scan rate of 10 mV s^{-1}, (b) GCD curves at a current density of 2 A $g^{-1}$of GO-M, and CSGO-M in various electrolytes; (c) specific capacitance (F g^{-1}) vs scan rate (V s^{-1}), and (d) specific capacitance (F g^{-1}) vs current density (A g^{-1} calculated from the CV curves)

4 Conclusion

This study presents a novel energy storage solution by synthesizing chitosan-modified graphene oxide binary metal composites through a hydrothermal process. The interaction between chitosan and graphene oxide enhances the surface area and facilitates even dispersion of binary metal oxides. Electrochemical assessments demonstrate superior performance, with the chitosan-modified composite exhibiting significantly higher specific capacitance compared to its counterparts. The addition of potassium ferrocyanide further enhanced its performance. The chitosan-modification of GO binary metal composite was able to address the stacking issues of graphene layers, thereby showing promise for advanced energy storage materials with improved efficiency across diverse applications.

References

1. Şahin, M.E., Blaabjerg, F., Sangwongwanich, A.: A comprehensive review on supercapacitor applications and developments. Energies **15**, 674 (2022)

2. Singh, A., Patil, U.Y., Khanra, P., Kumar, P., Kapoor, M.: Structural and stability profile of suramin-bound single-walled carbon nanotubes. AIP Conf. Proc. **2357**, 030011 (2022)
3. Peng, G., Qin, J., Volokh, M., Shalom, M.: Freestanding hierarchical carbon nitride/carbon-paper electrode as a photoelectrocatalyst for water splitting and dye degradation. ACS Appl. Mater. Interfaces **11**, 29139–29146 (2019)
4. Ravi Kumar, Y., Deshmukh, K., Kumar Sadasivuni, K., Khadheer Pasha, S.K.: Graphene quantum dot based materials for sensing, bio-imaging and energy storage applications: a review. RSC Adv. **10**, 23861–23898 (2020)
5. Fuertes, A.B., Sevilla, M.: Hierarchical microporous/mesoporous carbon nanosheets for high-performance supercapacitors. ACS Appl. Mater. Interfaces **7**, 4344–4353 (2015)
6. Oh, Y.J., et al.: Oxygen functional groups and electrochemical capacitive behavior of incompletely reduced graphene oxides as a thin-film electrode of supercapacitor. Electrochim. Acta **116**, 118–128 (2014)
7. Moitra, D., Anand, C., Ghosh, B.K., Chandel, M., Ghosh, N.N.: One-dimensional BiFeO3 nanowire-reduced graphene oxide nanocomposite as excellent supercapacitor electrode material. ACS Appl. Energy Mater. **1**, 464–474 (2018)
8. Ozkan, B.C., Soganci, T., Turhan, H., Ak, M.: Investigation of rGO and chitosan effects on optical and electrical properties of the conductive polymers for advanced applications. Electrochim. Acta **295**, 1044–1051 (2019)
9. El-Shafai, N.M., Shukry, M., Sharshir, S.W., Ramadan, M.S., Alhadhrami, A., El-Mehasseb, I.: Advanced applications of the nanohybrid membrane of chitosan/nickel oxide for photocatalytic, electro-biosensor, energy storage, and supercapacitors. J. Energy Storage. **50**, 104626 (2022)
10. Ma, H., et al.: Nickel cobalt hydroxide @reduced graphene oxide hybrid nanolayers for high performance asymmetric supercapacitors with remarkable cycling stability. ACS Appl. Mater. Interfaces **8**, 1992–2000 (2016)
11. Meera, M.S., Anil, A., Bhagya, T.C., Sameera, S.F., Remya, R., Shibli, S.M.A.: Development of cobalt oxide-decorated reduced graphene oxide-supported nickel alloy coating for efficient photocatalytic hydrogen generation. ACS Appl. Energy Mater. **6**, 10389–10403 (2023)
12. Liu, Y., et al.: Nanofabricated chitosan/graphene oxide electrodes for enhancing electrosorptive removal of U(VI) from aqueous solution. Sep. Purif. Technol. **290**, 120827 (2022)
13. Chen, Y., Liu, T., Zhang, L., Yu, J.: N-doped graphene framework supported nickel cobalt oxide as supercapacitor electrode with enhanced performance. Appl. Surf. Sci. **484**, 135–143 (2019)
14. Guo, Y., Sun, X., Liu, Y., Wang, W., Qiu, H., Gao, J.: One-pot preparation of reduced graphene oxide (RGO) or Au (Ag) nanoparticle-RGO hybrids using chitosan as a reducing and stabilizing agent and their use in methanol electrooxidation. Carbon **50**, 2513–2523 (2012)
15. Yu, M., et al.: Hydrothermal synthesis of $NiCo_2O_4$ nanowires/nitrogen-doped graphene for high-performance supercapacitor. Appl. Surf. Sci. **314**, 1000–1006 (2014)
16. Wei, X., Wang, C., Dou, P., Zheng, J., Cao, Z., Xu, X.: Synthesis of $NiCo_2O_4$ nanoneedle@polypyrrole arrays supported on 3D graphene electrode for high-performance detection of trace Pb2+. J. Mater. Sci. **52**, 3893–3905 (2017)
17. Ren, R., Zhong, Y., Ren, X., Fan, Y.: Chitosan-based oxygen-doped activated carbon/graphene composite for flexible supercapacitors. RSC Adv. **12**, 25807–25814 (2022)

Arm Angle Dependence of X-shaped Metamaterial Resonator in the X-band Regime

Soumadri Samanta[1], Gaurav Pal Singh[2], Prince Jain[3], Arun Kumar Singh[3], and Neha Sardana[2(✉)]

[1] Institute of Nano Science and Technology, Mohali 140306, India
[2] Indian Institute of Technology Ropar, Rupnagar 140001, India
nsardana@iitrpr.ac.in
[3] Punjab Engineering College (Deemed to be a University), Chandigarh 160012, India

Abstract. An angle dependent tunable design of an X-shaped metamaterial resonator (MMR) was theoretically and experimentally investigated in X-band (8–12 GHz). Single X-shaped MMR structures made of copper were simulated by changing the angle θ (w.r.t. magnetic field axis), and the data was confirmed experimentally with a Vector Network Analyser (VNA). The results indicated that with increasing angle, the resonance frequency was shifted towards the higher frequencies. Surface current data attributes the resonance to its magnetic component. Furthermore, the effective intrinsic parameters were calculated using Nicolson-Ross-Weir (NRW) technique to verify the metamaterial behavior of the resonator. The model with $\theta = 60°$ showed the best response, with a low return loss and a high figure of merit (FOM), which can be used for X-band stealth applications and radar communications.

Keywords: Metamaterials · Dielectric Resonator · X-Band · Resonance

1 Introduction

The metamaterial structuring has brought about a revolution in contemporary technology as it can exhibit unique electromagnetic properties and provide a platform for controlling electromagnetic waves. Metamaterials are artificial structures created by random or periodic inclusion of a structure, leading to properties unfounded by any material in nature. Metamaterials display negative electric permittivity (ε) and negative magnetic permeability (μ) [1], and can have tailorable electric [2] and magnetic [3] responses.

The dielectric resonator is an open resonating structure that is fabricated from low loss dielectric materials [4]. In recent years, resonators have received their due share of attention due to their small size, simplicity, low costs, stability, high radiation efficiency, relatively large bandwidth, low conductivity loss, and simple coupling schemes, which are often used in transmission lines. A resonator can operate over a significant portion of the electromagnetic spectrum by adjusting its geometrical length scale. Many research groups have demonstrated various applications of metamaterial resonators (MMRs),

A. Tiwari et al. (Eds.): iConFIRM 2023, SPPHY 308, pp. 183–190, 2024.
https://doi.org/10.1007/978-981-97-4557-9_19

especially cloaking [5], lenses beyond diffraction limit [6], wideband absorbers [7], filters [8] etc.

In our study, we computationally and experimentally investigated a single X-shaped dielectric resonator for absorber applications. X-shaped resonators can be designed to resonate at specific frequencies based on their dimensions and materials. This allows for precise tuning to target certain applications. Metamaterial behavior was observed for a large frequency range by tuning the arm angle. The planar structure of the proposed MMR is easy and cheap to fabricate and can be used in different applications like sensors, short-wave communication, radio and television receivers, etc.

2 Materials and Methods

The structure was designed in such a way that the FR4 dielectric substrate was sandwiched between two X-shaped resonators made up of Cu metal, as shown in Fig. 1. In addition, the symmetrical structure had polarization-independent characteristics. The FR4 substrate dimensions were $22.86 \times 10.16 \times 1.6\ \text{mm}^3$ and that of Cu metal X-shaped MMR inclusion was L1 = 1.7 mm, L2 = 9 mm and thickness 0.035 mm (Fig. 1a).

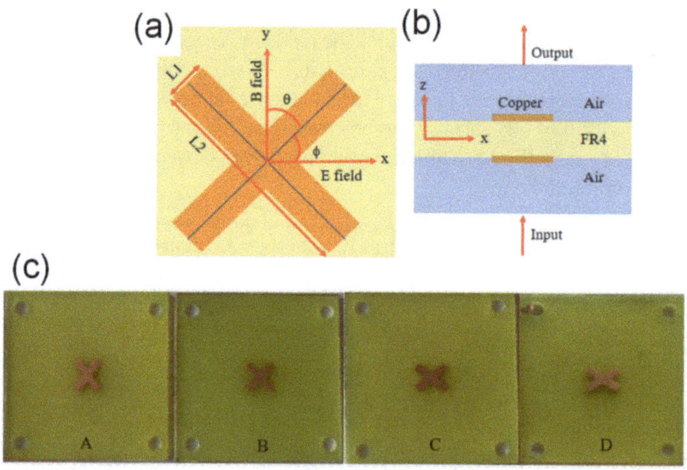

Fig. 1. Schematic showing (a) top view and (b) side view of X-shaped MMR. (c) Top view of the experimentally fabricated samples.

The angle between the central axis of the arm of the X shape and the applied magnetic field direction (y) was θ. Therefore, the angle w.r.t. electric field (x), $\phi = 90 - \theta$. These angles were varied in a step size of 7.5°. The angle θ, in samples A, B, C, and D was 37.5°, 45°, 52.5°, and 60°. The dielectric constant of the FR4 was 4.2 with a dielectric loss tangent of 0.02, and electric conductivity of copper was 5.8×10^7 S/m.

Using printed circuit board (PCB) manufacturing, A, B, C, and D models were fabricated, which is shown in Fig. 1(c). Experimental measurement of S parameters was performed in Keysight M9005A/M9374A vector network analyzer (VNA). A rectangular brass waveguide WR90 was used to hold the samples and carry out the experiment.

3 Simulation Methodology

In the simulations, the EM wave was considered to propagate along z-direction and was in transverse electric 10 (TE 10) mode, as shown in Fig. 1(b). The X-shaped MMR was optimized using COMSOL Multiphysics and MATLAB. Simulated reflection and transmission coefficients were used to retrieve effective medium parameters based on the Nicolson-Ross-Weir method [9]. Table 1 is the list of parameters which were used in the calculations.

Table 1. List of symbols used

Parameter	Symbol
Reflection coefficient	Γ
Relative impedance	Z
Relative permittivity	ε
Relative permittivity	μ
Thickness	d
Transmission coefficient	z
Angular frequency	ω
Frequency	f
Speed of light	c
S parameter: S11, Reflection coefficient	S11
S parameter: S21, Forward transmission coefficient	S21
Refractive index	n
Return loss	RL
Figure of merit	FOM

Applying this model, we considered a slab of material with relative permeability (μ), relative permittivity (ε), relative refractive index (n), and thickness (d). The reflection coefficient (Γ) of the wave incident at the interface between the slab and free space is expressed as [10]:

$$\Gamma = \frac{Z-1}{Z+1} \tag{1}$$

$$\text{where, the relative impedance } Z = \sqrt{\mu/\varepsilon} \tag{2}$$

For a finite value of d, the transmission coefficient (z) between two ends of the slab can be written as:

$$z = exp[-j(\omega/\mathrm{c})\sqrt{\mu\varepsilon}d] \tag{3}$$

where, c is the velocity of light and $\omega = 2\pi f$ is the angular frequency. Hence, the corresponding S parameters i.e., S11 and S21 can be expressed as:

$$S11(\omega) = \frac{(1 - z^2)\Gamma}{1 - \Gamma^2 z^2} \tag{4}$$

$$\mathrm{S21}(\omega) = \frac{(1 - z^2)z}{1 - \Gamma^2 z^2} \tag{5}$$

Further, the material properties can be calculated as follows:

$$\mu = \frac{2c(1 - S21 + S11)}{j\omega d(1 + S21 - S11)} \tag{6}$$

$$\varepsilon = \mu + j\frac{2cS11}{\omega d} \tag{7}$$

$$n = \sqrt{\mu\varepsilon} \tag{8}$$

The S parameter, S11 (reflection coefficient) represents the ratio of the reflected voltage wave to the incident voltage wave at a specific port of a device. The S parameter, S21 (forward transmission coefficient) represents the ratio of the transmitted voltage wave to the incident voltage wave at a specific port of a device. The S parameters are typically measured in decibels (dB) and are the pivotal parameters in deciding the spectral range of the device.

The parameters return loss (RL) and figure of merit (FOM) decide the overall performance of the device.

$$\mathrm{RL} = -20\, log_{10}|S11| \tag{9}$$

$$\mathrm{FOM} = -Real(n)/Imaginary(n) \tag{10}$$

Return loss (dB) is a measure of the amount of power reflected back from a device or component relative to the power incident upon it. Return loss quantifies how well a device or component matches the impedance of the system. FOM quantifies the trade-off between the ability of the material to transmit light and its propensity to absorb light; hence, it is critical in optimizing the performance of optical materials.

4 Results and Discussions

The S parameter values were simulated in COMSOL for frequency regime 8.0–12.0 GHz. The values of reflection coefficient (S11) are shown in Fig. 2a. It was observed that the resonance frequency shifts towards the higher frequencies, and the value of reflection coefficient decreased with the increase of the arm angle value (θ). Additionally, it was observed that the frequency gap between each consecutive dip position increases as the angle θ increases. The model D had the lowest S11 (Fig. 2a) and the highest S21 (supplementary material Fig. S1). Figure 2b shows that the experimental data measured with VNA matches well with the experimental results. The slight variations may be attributed to the losses in rectangular waveguide.

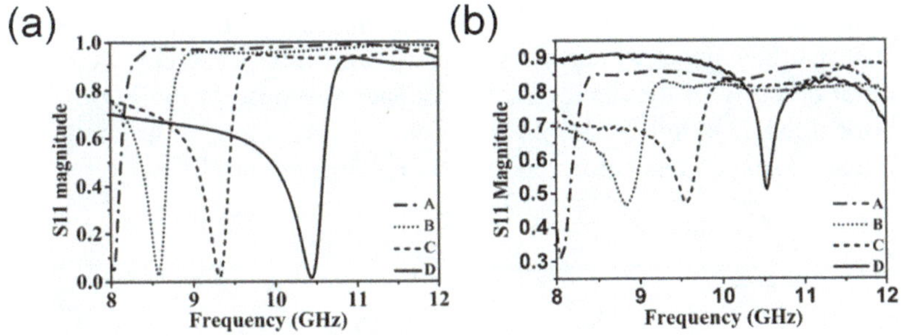

Fig. 2. a) Simulated and b) experimental frequency response of S11 magnitude of models A, B, C and D

The simulated surface current of model D is shown in Fig. 3. The surface current on the top layer and bottom layer of the resonator shows antiparallel currents at the resonance frequency. Similarly, the structures for model A, B, and C show similar antiparallel surface currents at their corresponding resonance frequencies. It is known that only for magnetic resonance, antiparallel surface currents are observed [12]. Furthermore, this magnetic resonance also explains the resonance frequency shift towards a higher frequency with increasing θ. The electric and magnetic fields were directed along the x and y axis, respectively (Fig. 1a). The resonance frequency is inversely proportional to the size of the resonating material. Hence, as θ increases, the spread along y axis decreases, and therefore, resonance shifts towards higher frequencies.

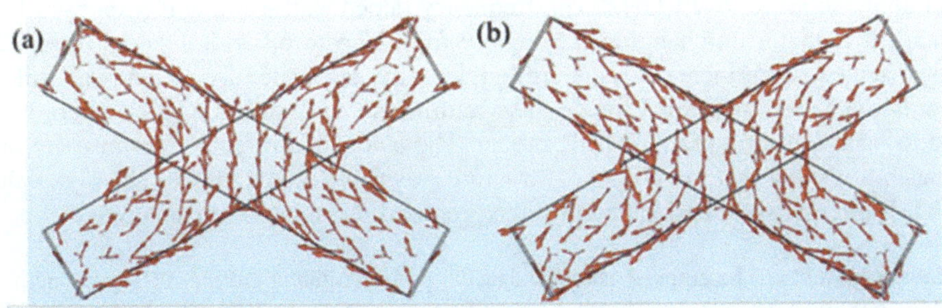

Fig. 3. Surface current of model D, (a) top layer and (b) bottom layer of Cu inclusion at the resonance frequency of 10.42 GHz

Figure 4a shows the simulated return loss and the value of intercept w.r.t. frequency at −10 dB. As predicted, the model D had the lowest return loss. The bandwidth and the size of an MMR are inversely proportional to the square root of the permittivity of the material [11]. Thus, it was confirmed that the bandwidth of the MMR was proportional to its size. This bandwidth increase can be attributed to the increase in the circular cross-section of the resonator with an increasing value of θ.

The material parameters of the MMRs are shown in supplementary material Fig. S2, which confirms their negative permittivity and permeability, thereby, conforming the metamaterial property of the device. The frequency response of the FOM is shown in Fig. 4b. Model D had a high FOM (25.7) which was 28.5% higher as compared with a state of the art H shaped resonator, reported by F. Dincer et al. [8].

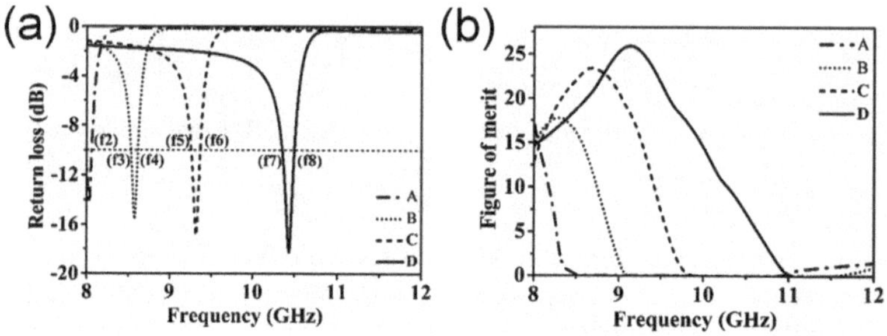

Fig. 4. Simulated frequency response of a) return loss and b) figure of merit of models A, B, C, and D

5 Conclusions

X-shaped MMRs having different arm angles were proposed and characterized for X band application i.e. 8–12 GHz. The frequency dependent arm angle response of an X shaped resonator on changing the angle θ from 37.5 to 60 with a step size of 7.5° was presented. With increasing the arm angle θ, the resonance frequency was shifted towards the higher frequency range. The return loss and scattering parameters were used to establish the optimal MMR model. The calculated effective permittivity and permeability show that the proposed structure was an effective MMR. The high value of FOM suggests good performance of the proposed metamaterial structure.

Acknowledgments. The authors acknowledge the project funding from SERB grant number ECR/2016/000150 for this work. Furthermore, SS acknowledges the JRF-SRF funding provided by Institute of Nano Science and Technology.

Supporting Information

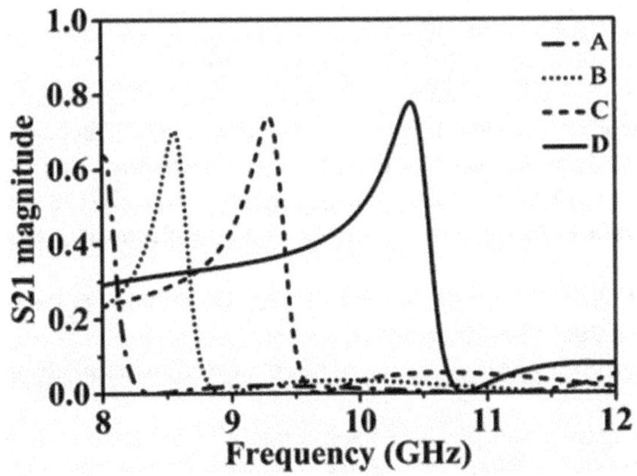

Fig. S1. Simulated frequency response of S21 magnitude of models A, B, C, and D

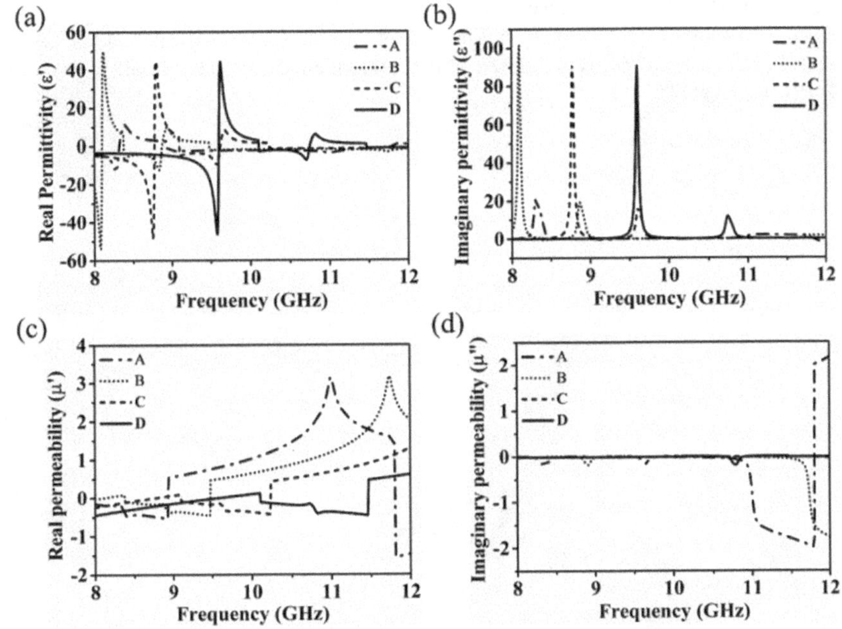

Fig. S2. Frequency response of (a) Real part of permittivity (ε′), (b) Imaginary part of permittivity (ε″), (c) Real part of permeability (μ′) and (d) Imaginary part of permeability (μ″) of models A, B, C and D

References

1. Smith, D.R., Padilla, W.J., Vier, D.C., et al.: Composite medium with simultaneously negative permeability and permittivity. Phys. Rev. Lett. **84**, 4184 (2000)
2. Pendry, J.B., Holden, A.J., Stewart, W.J., Youngs, I.: Extremely low frequency plasmons in metallic mesostructures. Phys. Rev. Lett. **76**, 4773 (1996)
3. Pendry, J.B., Holden, A.J., Robbins, D.J., Stewart, W.J.: Magnetism from conductors and enhanced nonlinear phenomena. IEEE Trans. Microw. Theory Tech. **47**, 2075 (1999)
4. Wiltshire, M.C.K., Pendry, J.B., Young, I.R., et al.: Microstructured magnetic materials for RF flux guides in magnetic resonance imaging. Science (80-) **291**, 849 (2001)
5. Schurig, D., Mock, J.J., Justice, B.J., et al.: Metamaterial electromagnetic cloak at microwave frequencies. Science **314**, 977 (2006)
6. Fang, N., Lee, H., Sun, C., Zhang, X.: Sub-diffraction-limited optical imaging with a silver superlens. Science **308**, 534 (2005)
7. Landy, N.I., Sajuyigbe, S., Mock, J.J., et al.: Perfect metamaterial absorber. Phys. Rev. Lett. **100**, 207402 (2008)
8. Dincer, F.: Investigation and physical interpretation of H-shaped metamaterials in X-band waveguide for microwave filter applications. J. Electron. Mater. **45**, 812 (2016)
9. Nicolson, A.M., Ross, G.F.: Measurement of the intrinsic properties of materials by time-domain techniques. IEEE Trans. Instrum. Meas. **19**, 377 (1970)
10. Domich, P.D., Baker Jarvis, J., Geyer, R.G.: Optimization techniques for permittivity and permeability determination. J. Res. Natl. Inst. Stand. Technol. **96**, 565 (1991)
11. Petosa, A.: Dielectric resonator antenna handbook. Artech House, Norwood, MA, USA (2007)
12. Zhou, J., Dong, J., Wang, B., et al.: Negative refractive index due to chirality. Phys. Rev. B **79**, 121104 (2009)

Author Index

A. Tiwari et al. (Eds.): iConFIRM 2023, SPPHY 308, p. 191, 2024.
https://doi.org/10.1007/978-981-97-4557-9

Kolibri 360 GmbH
Ferdinand-Jühlke-Straße 7
99095 Erfurt, Deutschland
produktsicherheit@kolibri360.de